DYNAMICS OF MOLECULE SURFACE INTERACTIONS

DYNAMICS OF MOLECULE SURFACE INTERACTIONS

Gert D. Billing

H. C. Ørsted Institute
University of Copenhagen
Denmark

A Wiley-Interscience Publication
JOHN WILEY & SONS, INC.
New York • Chichester • Weinheim • Brisbane • Singapore • Toronto

This book is printed on acid-free paper. ∞

Published simultaneously in Canada.

For customer and ordering service, call 1-800-CALL-WILEY.

Library of Congress Cataloging-in-Publication Data:

Billing, Gert D. (Gert Due)
Dynamics of molecule surface interactions / Gert D. Billing.
p. cm.
Includes index.
ISBN 0-471-33108-2 (alk. paper)
1. Gas-solid interfaces. 2. Surface chemistry. 3. Molecules--Surfaces. 4. Molecular dynamics. I. Title.
QD509.G37B55 2000 99-38866
541.3′3—dc21 CIP

Printed in the United States of America.

10 9 8 7 6 5 4 3 2 1

To Cathrine, Oliver, and Adam

CONTENTS

PREFACE

Dynamics of molecule-surface interaction is a new interdisciplinary field between chemistry and physics. Thus it is possible to take the chemist's or the physicist's point of view. The present monograph is a chemist's point of view. In order fully to understand the area it is necessary to have a feeling for the dynamics and reactivity of molecules, i.e., chemistry, but also the properties and physics of the solid are important issues. These latter subjects are usually only taught in physics courses. The complete area concerning chemistry at surfaces is immense and the present text does not attempt to cover the area in full. It is impossible to do so in a single book. It does, however, contain a description of aspects which recent research activity have proven to be of importance. One should also remember that the system under investigation is a many-body system. Therefore, methods which have been used to treat few-body problems will either have to be abandoned or at least drastically modified. Some of the approximations described in the book have proven useful in gas-phase chemistry. But the final methodology, in the area of molecule-surface dynamics, is not yet fully established. A contour of adequate methods is, however, visible and my purpose is to give the reader an overall view of the area, methods, and problems. The text will be slightly biased toward the author's personal experience and "points of view". But I believe that the area is covered reasonably well — at least this is so if the content of the references is included. The book tries to take a microscopic molecular dynamics point of view rather than a macroscopic kinetic point of view. The reason is that it is the trend in modern research to do so, experimentally as well as theoretically. If complicated processes such as the catalytic ones are to be fully understood it implies an understanding from first principles. Therefore, the molecular approach is obviously the most promising one. However, we do enter an area where the first principle equations are not directly solvable. The alternative is then to develop methods that for smaller systems have turned out to be adequate and at the same time have the advantage that they can be extended to larger systems. Such theories are, for instance, various approximate theories for the description of gas-phase dynamics by classical and semiclassical mechanics, boson and fermion descriptions of excitations in the solid, the density functional methods for the electronic structure problem, and so forth.

Obviously the book will be colored by personal experience and preferences. However, except maybe for Chapters 8 and 11, the selection of topics and examples has been taken from the literature in such a manner that, from a dynamical and computational point of view, most promising approaches have been included or mentioned.

The content of the book has been used in courses at the graduate as well as the undergraduate level for physics students interested in chemistry and vice versa. Parts of the book are rather elementary and other parts more advanced. A number of exercises have been added to the text. This is done in order for the student to get a better feeling for the various proofs and approximations which have been omitted in the text. The answer section gives a fairly complete solution of these exercises.

As mentioned, the book takes the molecular dynamical point of view. For students more interested in the kinetic or macroscopic point of view I would like to recommend the book by R. I. Masel, *Principles of Adsorption and Reaction on Solid Surfaces*, Wiley, 1996. However, the introductory chapter does include some of the kinetic concepts. Chapter 2 contains a number of the necessary basic concepts, such as Miller indices, lattice vectors, reciprocal lattice space, dispersion curves, lattice dynamics, and kinematics of atom-surface collisions. Chapter 3 discusses surface diffusion and Chapter 4 gives an overview of various suggestions for constructing protential energy surfaces. Quantum and wavepacket scattering on surfaces is described in Chapter 5. Various ways of including dissipative terms in the hamiltonian are also mentioned in this chapter. Dynamical methods for treating the motion of the surface and gas atoms are given in Chapters 6, 7, and 8. Chapter 6 is on classical mechanics, Chapter 7 introduces the generalized Langevin equation, and finally Chapter 8 describes a semiclassical method in which the surface atoms are treated as bosons. Chapter 9 mentions some electron gas theories and Chapter 10 the density functional theory. Finally, Chapter 11 considers electron-hole pair excitation, In the four Appendices we present the transition state theory, the valence bond scheme for one-electron atoms, the Hartree and Hartree-Fock methods for electron structure calculations, and finally the solution of the so-called inversion problem for obtaining the evolution operator for boson or fermion dynamics from the numerically integrated equations in an operator representation.

Finally, I would like to acknowledge students and colleaques for fruitful collaboration in this particular area, for suggestions and discussions concerning the presentation. Dr. Christian Mourier Laursen (The Danish Technical University) is acknowledged for providing me with Figures 4.2, 4.7, and 8.1. Professor John Avery (University of Copenhagen) is acknowledged for proofreading the manuscript.

GERT D. BILLING

1

INTRODUCTION

Chemical reactions at surfaces are important not only from a technological but also from a fundamental scientific point of view. The surface modifies in an important fashion the interaction between atoms in the gas phase. This modification depends on the electronic structure of the solid and on the spacial structure or geometrical pattern of the surface. By adding ad-atoms and changing the structure of the surface one has in principle an infinite number of ways of changing the reactivity at a surface. Thus, there are several reasons for understanding, at the molecular level, the many possibilities. In short, the study of the dynamics of molecule surface interaction is important for the understanding of chemical reactions and processes (catalyses, corrosion, epitaxial growth, etc.) at surfaces from a molecular point of view.

Although the dynamical problem of the reacting molecular system is of the same complexity as that encountered in gas-phase reaction dynamics, the presence of a surface adds additional processes and phenomena. Such phenomena are, for instance, not only the importance of the structure, including corrugation, steps, and surface anomalities, but also the interaction with the possible excitation processes in the solid, such as phonon (surface vibrations) and electronic excitations. Also for charge transfer and other nonadiabatic electronic processes in the gas phase, the importance of the surface temperature adds additional features to the problem. Aside from this, the various processes of interest occur on different time scales, from fast reactive chemisorption processes on the sub-pico second time scale to the relatively slow diffusion and desorption processes. Thus different theoretical tools are needed in order to describe the variety of processes and the large time span one needs to cover. Also the many-body problem of the solid combined with the few-body gas-phase problem makes it necessary to introduce different methods for treating the dynamics, from classical trajectories and

TABLE 1.1 Important Dynamical Aspects in Molecule-Surface Interaction

Process/Phenomenon
Corrugation, steps, dislocations, shape, structure
Surface temperature
Physisorption
Chemisorption
Initial excitation of the molecule
Phonon excitations
Electron-hole pair excitation
Nonadiabatic processes
Diffusion
Desorption

wavepacket propagation technique to the use of second quantization concepts. Table 1.1 gives an illustration of the processes of interest.

- The corrugation, i.e., the roughness of the surface, whether the surface has anomalities or not, can locally change the reactivity at a given surface site. STM (scanning tunneling microscopy) has revealed new information on the surface structure at the atomic level [1], and since the surface structure can be important for the molecular processes it is of course desirable also to be able to model this aspect theoretically. Thus the influence of surface defects, dislocations, and other unusual structure on catalytic rates is an important issue.

  The possibility of creating arrays of nanoparticle structures with a size of a few hundred Å has added to the importance of studying the effect of size and shape on chemical processes taking place on the "mesoscopic" scale [2, 3]. Certain molecular structures, such as zeolites, have pores which have the right size for supporting the transition state of a reaction and thereby changing the reactivity. In any case the structural effect on chemical reaction dynamics is an important one in surface chemistry as well as in other disciplines, such as biochemistry.
- The effect of changing the surface temperature is of importance for many catalytic reactions. However, when investigated at the molecular level, the effect of changing the surface temperature can be more subtle than normally is believed [6].
- The incoming molecules may be trapped in a physisorption (weakly bound) state at the surface. This happens if the energy of the atom/molecule is efficiently dissipated to the surface. Physisorption energies are typically less than about 20 kJ/mole.
- However, the energy may also be used to surpass a reaction barrier and the molecule may enter into a chemisorbed state. In this state the atoms are

strongly bound (often with several eV) to certain sites at the surface. Here it should be noticed that the reaction barrier is not as easily defined a concept as in gas-phase chemistry. The barrier depends, first, upon the site at which the molecule approaches. Second, the interaction between a molecule and the surface makes the barrier a dynamical one and hence comparison with experimentally determined activation energies and calculations requires a full molecular dynamics simulation of the problem.

- The process which happens at the surface is strongly influenced by the way the energy is distributed in the molecule. Thus it may in certain cases be more effective for surface reactions to occur, to distribute energy in the internal, i.e., rotational and vibrational, rather than in the translational degrees of freedom.
- Once the molecule is adsorbed at the surface it may move (diffuse) to another site which has different reactivity. Thus one important distinction which can be made between surface processes is whether reaction happens directly or after a motion or diffusion along the surface.
- During the interaction with the surface the molecule will excite various surface modes, such as surface vibrations (phonons) and electrons.
- By nonadiabatic processes we understand such processes as involve more electronic states relevant for the dynamics of the gas-phase molecules. These could be either electronic states, which are associated with the molecule asymptotically (far from the surface), or those of the surface electrons. However, when the molecule approaches the surface it is problematic to maintain this distinction.
- Finally, when the initial reaction or direct dynamical process in question has occurred, the products may stick to the surface or desorb to the gas phase. Thus the desorption mechanism is yet another important process from a fundamental point of view.

1.1 SIMPLE ADSORPTION MODELS

A quantity which is important from a catalytic point of view is the amount of gas which can be adsorbed at the surface. This is expressed in terms of the so-called adsorption isotherms. Langmuir [4] assumed that the adsorption rate was proportional to the fraction of free surface sites $1-\theta$ and the desorption rate proportional to the fraction of covered surface θ, and he obtained the expression

$$\theta = \frac{K[A]}{1+K[A]} \tag{1.1}$$

where $[A]$ is the concentration in the gas phase and $K = k_{ads}/k_{des}$ the ratio between rate constants for the adsorption and desorption processes. The kinetic data of catalytic reactions can most often be fitted to the Langmuir or the Langmuir-Hinshelwood type rate equation [5] (see below).

Introducing the rate of adsorption as the integrated flux of molecules A in the direction toward the surface with energy sufficient enough to overcome the adsorption barrier E_{ads}, we have

$$\text{Rate of adsorption} = [A](1-\theta)\int_{v_0}^{\infty} f(v_z)v_z p_{ads} dv_z \tag{1.2}$$

where $[A]$ is the number of molecules per unit volume, p_{ads} the sticking probability, $f(v_z)$ the one-dimensional velocity distribution function

$$f(v_z) = \sqrt{\frac{m}{2\pi k_B T}} \exp\left(-\frac{1}{2}mv_z^2/k_B T\right) \tag{1.3}$$

m the mass of the molecule, k_B Boltzmann's constant, and v_z the velocity in the z direction. The threshold velocity is defined by $E_{ads} = (1/2)mv_0^2$. If we assume that the sticking probability is independent of v_z we can easily carry out the integral and get

$$\text{Rate of adsorption} = [A]\sqrt{\frac{k_B T}{2\pi m}} \exp(-E_{ads}/k_B T)(1-\theta)p_{ads} \tag{1.4}$$

We also introduce the desorption rate as

$$\text{Rate of desorption} = k_{des}\theta \exp(-E_{des}/k_B T) \tag{1.5}$$

where k_{des} is a desorption rate coefficient and E_{des} the activation barrier for desorption (see Fig. 1.1). At equilibrium the two rates are the same and we get

$$[A] = \frac{\theta}{K(1-\theta)} \tag{1.6}$$

This expression is identical to the Langmuir expression (1.1). The constant K is given as

$$K = k_{ads}/k_{des} \tag{1.7}$$

where

$$k_{ads} = \sqrt{\frac{k_B T}{2\pi m}} p_{ads} \exp(E_b/k_B T) \tag{1.8}$$

and $E_b = E_{des} - E_{ads}$ is the binding energy (see Fig. 1.1).

The BET (Brunauer, Emmett, Teller) [7] isotherm gives an expression for the amount of gas adsorped as a monolayer V_m as a function of the pressure P of the gas molecules. The expression is

$$\frac{P}{V_{ads}(P_0 - P)} = \frac{1}{CV_m} + \frac{C-1}{CV_m}\frac{P}{P_0} \tag{1.9}$$

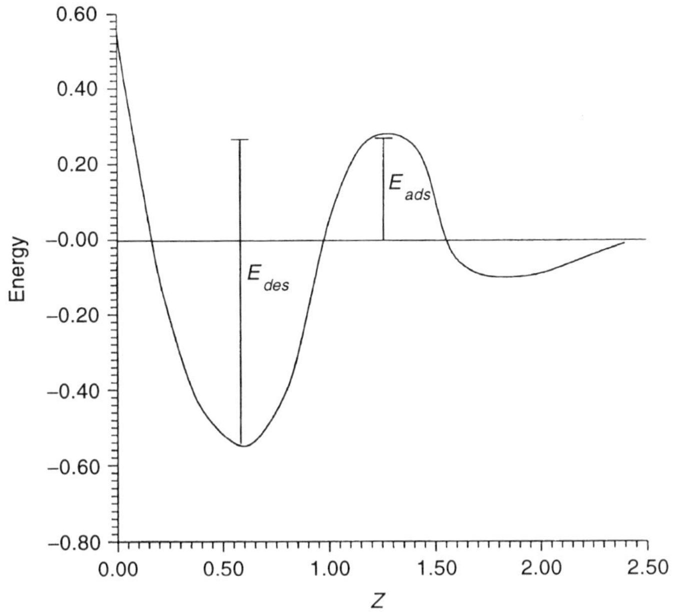

FIGURE 1.1 Schematic representation of adsorption/desorption energies. A molecule adsorbs with an activation barrier E_{ads} and desorbs with the barrier E_{des}. Z is the distance from the surface in Å and energies in 100 kJ/mol.

where V_{ads} is the volume of the adsorbed gas, P_0 the pressure of the saturated gas at a given temperature, and C a constant (for a derivation of the isotherm see, e.g., [130]). Since the molecules in the first layer are likely to chemisorb whereas molecules adsorbed on a monolayer of atoms are physisorbed, it is relevant to know how much of a given volume of gas can be adsorbed as a monolayer. The BET-isotherm gives this information as a function of gas pressure at a given temperature.

The adsorption process considered above has not involved any detailed knowledge on how the process occurs at the atomic scale. This is typical for the so-called "kinetic" point of view. We assume that a certain number of steps are relevant for the overall process and introduce (often unknown) rates for these processes. If the kinetic scheme fits the mascroscopic data we are satisfied. Contrary to this point of view is the dynamical one, where one wishes to understand the details of the molecular events. It will be the latter point of view which will dominate the remaining part of this book.

Often the surface reactions are referred to as one of two types. The Langmuir-Hinshelwood (L-H) type involves the reaction of two adsorbed molecules

$$A^* + B^* \rightarrow AB \tag{1.10}$$

where the * indicates surface-bound species. Consider two adsorbed molecules A and B located in neighbor wells at the surface and separated by a barrier. The vibrational motion in the well will take place with a frequency of about 10^{12} to 10^{13} times per second. For each vibration there is a probability for reaction with the adjacent molecule (eventually as a result of tunneling through a barrier). Thus the frequency is often called the attempt frequency, i.e., it is the number of times per time unit that the molecule or atom attempts to jump to a neighbor site. The reaction rate in the L-H mechanism is proportional to the fraction (θ_A, θ_B) of the surface covered with A and B molecules, i.e.,

$$v_{react}^{LH} = k\theta_A\theta_B \tag{1.11}$$

Introducing now the assumption (due to Langmuir) that the adsorption rate is proportional to the amount of free surface we have

$$v_{ads}^{A} = k_{ads}^{A}[A](1 - \theta_A - \theta_B) \tag{1.12}$$

$$v_{ads}^{B} = k_{ads}^{B}[B](1 - \theta_A - \theta_B) \tag{1.13}$$

where $[A]$ and $[B]$ are the gas phase concentrations. k_{ads}^{A} and k_{ads}^{B} are rate constants for the adsorption process.

For the desorption rates we have

$$v_{des}^{A} = k_{des}^{A}\theta_A \tag{1.14}$$

$$v_{des}^{B} = k_{des}^{B}\theta_B \tag{1.15}$$

Solving for θ_A and θ_B we get

$$v_{react}^{LH} = \frac{kK_AK_B[A][B]}{(1 + K_A[A] + K_B[B])^2} \tag{1.16}$$

where we have introduced the equilibrium constants $K_p = k_{ads}^{p}/k_{des}^{p} (p = A, B)$.

This reaction rate, obtained from the Langmuir-Hinshelwood model, should be compared with that for the direct reaction through the so-called Eley-Rideal mechanism, where a molecule in the gas phase reacts with a surface-bound one, i.e.,

$$A + B^* \rightarrow AB \tag{1.17}$$

The reaction rate is here proportional to the surface coverage of B molecules and the gas-phase concentration of A molecules, i.e., using the expression for θ_B just calculated we get

$$k_{react}^{ER} = k[A]\theta_B = \frac{kK_B[A][B]}{1 + K_A[A] + K_B[B]} \tag{1.18}$$

Under kinetic conditions the L-H mechanism dominates [19] (see also exercise 1.8). If the energy exchange with the substrate by excitation of phonons and/or electrons is efficient, then the L-H reaction will occur under kinetic conditions, and the rate constant for the reaction could for instance be estimated using transition state theory (see Appendix A) with Boltzmann distributed energies according to the given surface temperature. However, in the direct E-R reaction the incoming gas atom or molecule need not have a kinetic energy according to a Boltzmann distribution and since the energy exchange with the substrate is smaller, most of the available energy can be used to overcome a reaction barrier. Such conditions can be reached using molecular beam techniques and hence the direct mechanism is favored. These new experimental techniques have made necessary a shift from the kinetic and thermodynamical point of view to the molecular and dynamical. Aside from these two mechanisms a third type of mechanism has also been proposed, namely the precursor mediated mechanism. Here an adsorbed molecule reacts with a precusor, which is thought of as a rather mobile atom/molecule which can move at the surface until it encounters an adsorbed molecule and reacts. In practice a given reaction can be of one of these classes or an intermediate case. The reaction may even switch from one mechanism to another as a function of surface temperature [8].

If the interaction with the surface is strong, the surface-bound molecule will be trapped near the bottom of the absorption well. The barrier for jumping to the neighbor well is then high and may therefore be higher than the barrier for the direct reaction. Thus the important aspects here are the steps involving adsorption, diffusion, reaction at the surface, and desorption.

The understanding of the importance of the various features is presently incomplete. This means that the introduction of molecular concepts and molecular dynamical simulations of processes important for the understanding of for example, catalyses from first principles still has a long way to go. But a contour of adequate theoretical tools is beginning to appear. Therefore this monograph will try to describe some of the more successful methods in enough detail for the implementation of the methods in actual programming.

At the qualitative level we can understand many of the phenomena in terms of simple pictures [9]. Thus Figure 1.2 shows schematically the molecule-surface interaction potentials as functions of the distance z from the surface, as suggested by Lennard-Jones [9]. Lennard-Jones obtained the following simple interaction potential depending only on the distance parameter z

$$V_{int} = \frac{D}{2}\left[\left(\frac{z_0}{z}\right)^9 - 3\left(\frac{z_0}{z}\right)^3\right] \tag{1.19}$$

where D is the well depth obtained for $z = z_0$. In Chapter 4 we shall discuss this and other models in further detail.

The molecule interacts only weakly with the surface when trapped in a physisorption well. At low energies the molecule may be trapped in this well before it dissociates. Such a temporary state is called a precursor state, and the precursor mediated dissociation is expected to exhibit a decrease in dissociative sticking with increasing surface temperature. Higher surface temperature will tend to lower

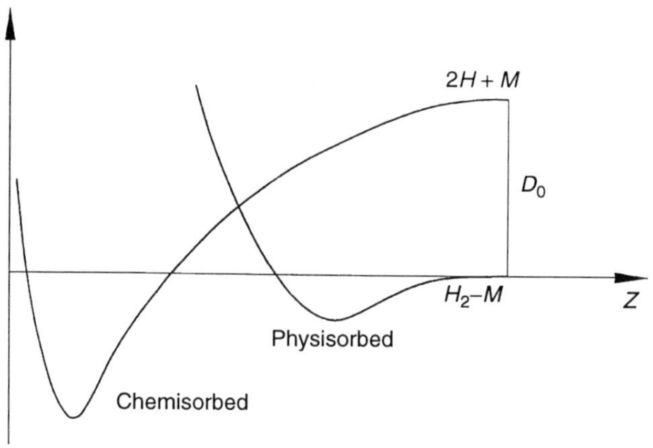

FIGURE 1.2 Schematic interaction between a molecule or two atoms and a metal surface M. D_0 denotes the gasphase dissociation energy. A barrier for dissociative chemisorption is seen where the physisorption (the molecule-metal) and the atom-metal potentials cross.

the probability for forming a precursor state. Although we often just encounter a molecular physisorption and a dissociative chemisorption well, cases do exist where the molecule can be found in an intermediate molecular chemisorption well [14].

The energy needed to dissociate the molecule is in the gas phase D_0. It lies typically in the range from a few eV to about 9 eV for a molecule such as nitrogen. Due to a possible charge transfer process from the surface to the molecule, filling electrons in the antibonding orbitals, the molecular bond is weakened. The interaction between the electrons in the solid and the molecule A increases the affinity for forming a negative ion A^-, and when the affinity equals the work function, i.e., the energy needed to remove an electron from the surface, the charge transfer becomes a natural process. In this sense the mechanism is similar to the one known from the harpoon model [10] in gas-phase reactions, where the reaction

$$M + X_2 \rightarrow MX + X \tag{1.20}$$

between the metal atom M and halogen molecule X_2 takes place through the steps

$$M \rightarrow M^+ + e^- \tag{1.21}$$

$$X_2 + e^- \rightarrow X_2^- \tag{1.22}$$

and finally

$$M^+ + X_2^- \rightarrow MX + X \tag{1.23}$$

The difference between the ionization energy (I) and the electron affinity (A) depends on the distance between the metal atom and the halogen molecule, and hence the electron can jump from M to X_2 when I − A ∼ 0.

The molecule surface interaction can occur in a similar fashion, especially for metals where the work function (ionization energy) is relatively small (a few eV). The interaction between two atoms and the surface as well as the molecule and the surface can be visualized by the curves on Figure 1.2. The two curves cross at a certain distance and the energy at that point is the activation barrier for dissociative chemisorption. At large distances a charged particle interacts with the surface through a classical image potential

$$V_{im} = -\frac{q^2}{4(z - z_{im})} \tag{1.24}$$

where q is the charge and z_{im} the position of the image plane. The factor of 4 in the denominator can be explained by noting that the distance to the image charge is $2(z - z_{im})$ and since it is an induced charge we get an additional factor of 2. As a consequence of this interaction, we can imagine that the affinity level and ionization potentials vary with the distance approximately as

$$\mathrm{A}(z) = \mathrm{A} + \frac{e^2}{4(z - z_{im})} \tag{1.25}$$

$$\mathrm{I}(z) = \mathrm{I} - \frac{e^2}{4(z - z_{im})} \tag{1.26}$$

bringing about the crossing at some value for z.

Since the complete potential energy surface for a diatomic molecule and the solid depends on the corrugation of the surface and of all six variables needed for describing the position of the two atoms, the above picture is, of course, too simplistic.

Also the energy transfer to the surface by inelastic processes is not included in this simplified picture. Figure 1.3 illustrates that the incoming atom can by the collision induce inelastic excitations in the surface. These inelastic excitations are represented by excitation of the oscillators to the left. The inelastic excitation makes the molecule or atom lose energy, such that it may be trapped in the physisorption well. This energy accommodation has also been described using simple models. These have treated the atoms as hard spheres (billiard balls). Here the energy transfer depends on the mass ratio between the incoming atom m_g and the surface atoms m_s. The energy transfer also depends strongly on the approach angle θ being large for perpendicular approach $\theta = 0$ and approaching zero for glancing collisions $\theta = \pi/2$. Since the kinetic energy in the direction toward the surface is proportional to $\cos^2\theta$, we would expect that also the energy accommodation is proportional to $\cos^2\theta$. Actually a simple model due to Baule [11] gives the following expression for the energy accommodation coefficient

$$\alpha = \frac{4\mu\cos^2\theta}{(\mu + 1)^2} \tag{1.27}$$

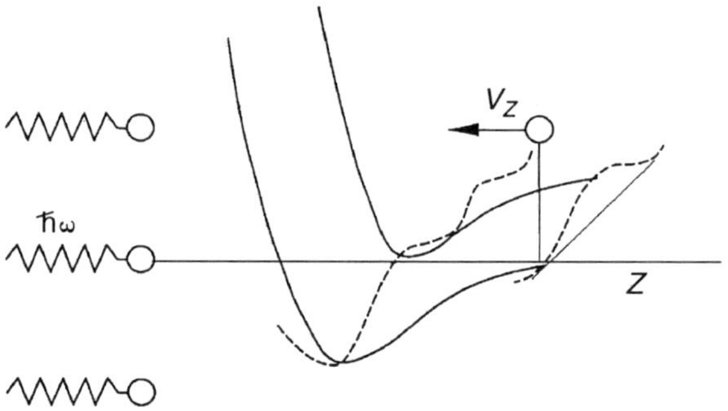

FIGURE 1.3 Simplistic representation of the surface inelastic processes represented by excitation of the oscillators with energies $\hbar\omega$.

where $\mu = m_g/m_s$. Although simple but more realistic models do exist [12] it is the purpose of the present monograph to present methods where the energy accommodation is calculated numerically. The energy accommodation is responsible for the possibility that the molecule may be trapped in the physisorption well. Once in the physisorption well it can diffuse along the surface. The diffusion process can eventually bring the molecule to a "hot spot" at the surface where it may desorb due to phonon-assisted desorption, or it may move to a place where the barrier for chemisorption is lower, and hence it may dissociate. Thus although simple pictures may in principle be drawn of these processes, the quantitative treatment requires an approach which treats all of the possible processes in detail, i.e., solves the dynamical equations for the many-body problem of the nuclei and electrons involved.

EXERCISES

1.1: Derive the Langmuir isotherm.

1.2: Determine the constant K in the Langmuir isotherm from the following adsorption data. The adsorbed volume of gas per m^2 surface refer to 298 K and 1 atm condition. Assume a cubic lattice with lattice constant $a = 2.5$ Å and that the gas is adsorbed as a monolayer with one molecule/unit cell (with area a^2).

Gas Pressure	Adsorbed Volume Gas/m^2
0.0052 atm	0.016 cm^3
0.017 atm	0.050 cm^3
0.045 atm	0.116 cm^3

1.3: Show that the Langmuir adsorption isotherm is modified to

$$\theta = \frac{K\sqrt{P}}{1 + K\sqrt{P}} \tag{1.28}$$

if the molecule completely dissociates at the surface. P is the gas pressure.

1.4: Determine the best value of K in the Langmuir isotherm which fits the following adsorption data for hydrogen at 362 K on Ni(110) (data from [13]). Notice that hydrogen dissociates on Ni(110).

Pressure (torr)	θ
$2\ 10^{-7}$	0.041
$3\ 10^{-7}$	0.059
$4\ 10^{-7}$	0.171
$5\ 10^{-7}$	0.482
$1\ 10^{-6}$	0.781
$4\ 10^{-6}$	0.890
$2\ 10^{-5}$	0.984

1.5: Calculate the rate constant at 300 and 1000 K for dissociative sticking of hydrogen on a Cu(001) surface using transition state theory (see Appendix A). The moment of inertia is $I = 0.227$ amuÅ^2. The vibrational frequency is 4300 cm^{-1}. At the transition state we find the following frequencies: i 319, 81.6, 400, 731, 791, and 1282 cm^{-1}. The activation barrier is $E_0 = 40$ kJ/mol.

1.6: Assume that the mirror image distance is $z_{im} = 0$ an affinity of 0.5 eV and I = 3 eV. Calculate the distance z^* where the electron transfer is a natural process.

1.7: Calculate the angle-averaged energy accommodation coefficient $\langle\alpha\rangle$ using the Baule formula (1.27).

1.8: Derive an expression for $k(T)_{ER}/k(T)_{LH}$, i.e., the ratio of the rate constants for the Eley-Rideal and the Langmuir-Hinshelwood rates for reaction of two atoms $A + A \rightarrow A_2$. Assume the atom mass 1 amu, a temperature of 400 K, and vibrational frequencies 100 cm^{-1} for the surface-bound atom. The gas pressure is 1 torr and the coverage $\theta_A = 1$ atom/site. At what binding energy of the atom is the ratio equal to unity? *Hint*: Use the transition state approach (see Appendix A).

1.9: Calculate the number of hydrogen molecules striking a 1-cm^2 area of a surface each second if the gas pressure is 10^{-8} atm and the temperature 300 K. With one site per unit cell (6.25 Å^2), how long does it take to cover an area of 1 cm^2?

2

BASIC CONCEPTS

2.1 LATTICE STRUCTURE AND MILLER INDICES

The coordinate system we will use to describe the molecule surface system has its z-axis perpendicular to the surface, i.e., in the direction of the surface normal (see Fig. 2.1). The crystal shape is determined by the unit cell, which is defined by the values of a, b, and c specifying the length along three axes $\mathbf{a}$, $\mathbf{b}$, and $\mathbf{c}$. For the simple lattices considered here the axes are perpendicular to each other and have therefore been denoted $\mathbf{x}$, $\mathbf{y}$, and $\mathbf{z}$. In order to characterize the directions and atomic planes in the crystal one introduces various notations.

Here we shall just mention the so-called Miller indices, which will be denoted (h, k, l). These three indices indicate the intercepts of a plane with the crystal axes. Additional indices are needed in noncubic lattices and yet another notation $\langle ijk \rangle$ is used to characterize the direction in a lattice. For a discussion of this the reader is referred to more specialized literature.

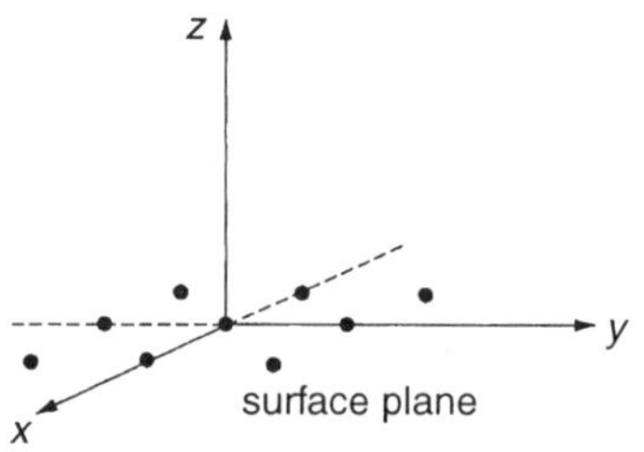

FIGURE 2.1 The coordinate system used to describe the dynamics is placed with the z-axis perpendicular to the surface.

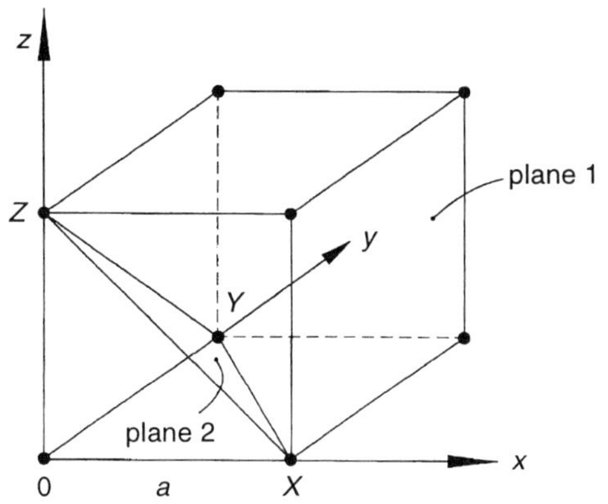

FIGURE 2.2 The planes (100) (plane 1) and (111) (plane 2) for a simple cubic crystal.

Returning to the Miller indices, consider, for example, a simple cubic lattice (Fig. 2.2); we see that plane 1 cuts the x, y and z-axes at a, ∞ and ∞ respectively, whereas plane 2 cuts the axes at a, b and c. Thus the Miller indices (h, k, l) [15] are defined as:

$$h = a/Ox = 1 \text{ (plane 1) and } 1 \text{ (plane 2)}$$

$$k = b/Oy = 0 \text{ (plane 1) and } 1 \text{ (plane 2)}$$

$$l = c/Oz = 0 \text{ (plane 1) and } 1 \text{ (plane 2)}$$

for the two cases. Here Ox denotes the distance at which the x-axis is cut by the plane, and so on. Furthermore, we have to specify the basic structure of the lattice. In 1848 it was shown by Bravais [16] that all lattices could be assigned to one of the 14 so-called Bravais-lattices, arranged in 7 classes of which the orthorombic and cubic are among the most common. Metal structures are most often cubic lattices (simple, face, or body-centered) (Fig. 2.3) or a close-packed hexagonal lattice. Thus metals such as copper, silver, gold, platinum, nickel, and palladium are all faces-centered cubic lattices at room temperature.

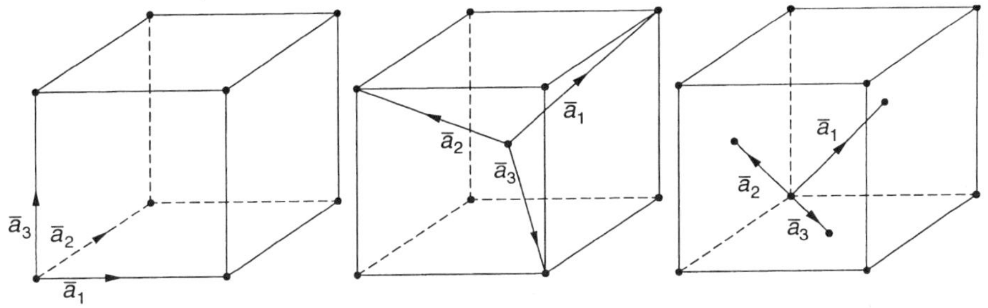

FIGURE 2.3 Simple cubic lattices (simple, body, and face-centered) and basis vectors.

The close-packed hexagonal structure is similar to the simple cubic—just with a distance $b \neq a$ between the layers.

The crystal structure and the way it is cut influences the density of atoms in each plane as well as the distance between the planes. Thus the perpendicular distance d between two adjacent planes in the lattice can be expressed in terms on the lattice constants (a, b, c) and the Miller indices. For the orthorombic unit cells we have

$$d^{-1} = \sqrt{(h/a)^2 + (k/b)^2 + (l/c)^2} \tag{2.1}$$

which for the cubic lattice with $a = b = c$ gives

$$d = \frac{a}{\sqrt{h^2 + k^2 + l^2}} \tag{2.2}$$

These formulas are special cases of the general expression considered in exercise 2.1.

The structure of the lattices can be determined experimentally using X-ray diffraction experiments. Early experiments were carried out by Bragg in 1912 on the NaCl lattice [17]. These experiments revealed that sodium chloride was a simple cubic lattice. The so-called Bragg condition for reflection of the X-ray can be understood from Fig. 2.4. The X-ray beam will be scattered from the

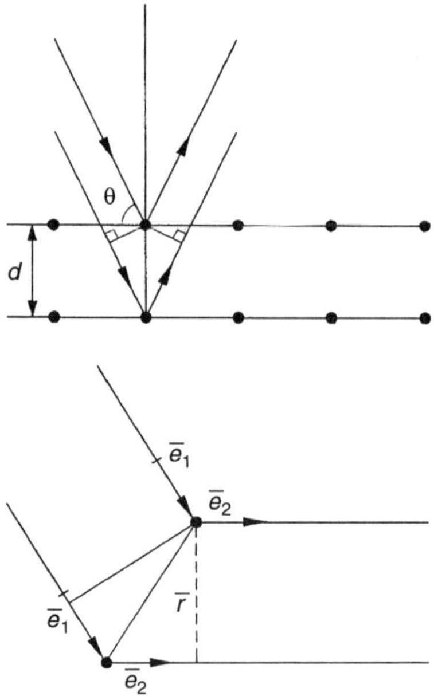

FIGURE 2.4 The condition for constructive Bragg-interference. d is the distance between the two layers and θ the incident angle. The vectors $\mathbf{e}_1$ and $\mathbf{e}_2$ are unit vectors in the directions of the incident and scattered beams respectively.

different layers of atoms. Only under certain conditions will the scattered waves be in phase, i.e., interfere constructively with each other. The condition is that the extra distance traveled by a beam scattered from the second layer, and so on, is an integral number times the wavelength, i.e.,

$$n\lambda = 2d \sin\theta \tag{2.3}$$

where θ is the incident angle, d the spacing between the lattice planes, n an integer, and λ the wavelength of the X-ray beam (see Fig. 2.4). From the knowledge of θ and the wavelength is it possible to determine d. Thus the various crystal structures have a specific scattering pattern according to the planes in the crystal which reflect the X-ray beam. A simple cubic lattice has scattering from the (100), (110), (111), (200) ... planes, the BCC crystal from (110), (200), (211), (220) ... planes, and the FCC lattice from (111), (200), (220), and (311) .... From eqs. (2.2) and (2.3) we may calculate these reflection angles and from the reflection angles the structure and lattice constant can be determined (see also exercise 2.5).

The Bragg condition can be derived on more general grounds by considering scattering from two points separated by a vector $\mathbf{r}$. Consider two unit vectors $\mathbf{e}_1$ and $\mathbf{e}_2$ in the directions of the incident and scattered beams from the two points (see Fig. 2.4). The condition for constructive interference is then

$$\mathbf{r} \cdot (\mathbf{e}_2 - \mathbf{e}_1) = n\lambda \tag{2.4}$$

and since the distance $\mathbf{r}$ between two lattice points can be written in terms of the lattice basis vectors $\mathbf{a}_i$ as $\mathbf{r} = \Sigma_i m_i \mathbf{a}_i$ (where m_i are integers) we get $\mathbf{a}_i \cdot (\mathbf{e}_2 - \mathbf{e}_1) = h, k,$ and $l\lambda$ for $i = 1, 2,$ and 3 respectively.

These conditions are the so-called Laue equations from which the X-ray diffraction patterns as well as the Bragg relation can be derived (see exercise 2.3).

- As mentioned in Chapter 1, the ability for a surface to act as a catalyst depends upon the electronic structure of the array of atoms with which the incoming molecule interacts. More specifically if we increase the number of d-electrons in the metal we get a drop in binding energy between the chemisorbed atom and the metal—however the activation barrier is also increased. Thus the first effect increases the reactivity, the second lowers it. These two opposing effects make metals with about half-filled d-levels the best catalysts. If one therefore plots the reactivity as a function of the number of d-electrons over the periodic table one obtains the so-called "volcano curves" with a maximum for metals such as Fe and Ru. At least this is so for the ammonia syntheses, where the rate-determining step is the dissociation of nitrogen. However, the structure of the surface is also known to have influence on its catalytic effects. Thus open and more rough surfaces as the (111) surface are more reactive than closed-packed (110) surface. This is quantified for the ammonia synthesis in Table 2.1. However, for other reactions, such as the methanol synthesis, the opposite trend is found [275]. Table 2.1 shows the reaction rates for the ammonia synthesis over an iron surface, i.e., for the reaction

$$N_2 + 3H_2 \rightarrow 2NH_3 \tag{2.5}$$

- The dissociation of nitrogen is the rate-determining step and the iron surface diminishes the energy needed to break the bond from about 9 eV in the gas phase to about 0.5 eV at the surface. Thus if a simple Arrhenius theory for the rate constant can be assumed

$$k(T) = A \exp(-E_a/k_B T) \tag{2.6}$$

where E_a is the activation energy and k_B the Boltzmann constant, we get at 700 K (the reactor temperature) an enhancement of the reaction by over sixty orders of magnitude. However, this analysis is too simple since the adsorbed nitrogen atoms are strongly surface bound. Hence energy is needed to remove the atoms from the surface if the reaction with them had to take place in the gas phase. But by letting the nitrogen and hydrogen atoms react at the surface, we can, in a consecutive manner, form ammonia, which is only weakly surface bound (see Fig. 2.5).

In calculations as well as in experiments one is therefore interested in studying the activity of a certain surface structure. In order to obtain the desired surface structure it may then be necessary to rotate and cut the crystal.

Most metals have, as mentioned, a crystal structure which is either a body-centered cubic, a face-centered cubic, or a closed-packed hexagonal structure. As an example consider the Pt(111) surface. The basic structure of platinum is an FCC (face-centered cubic) lattice. If we wish to cut the surface along the 111-plane and to find the (x, y, z) positions of the atoms in layer 1, 2, and so on such that the z-axis is perpendicular to the 111-plane, this can be done by rotating the

TABLE 2.1 Reactivity of Various Iron Surfaces for the Formation of Ammonia

Surface	Reactivity
111	13
211	9.8
100	1.9
210	1.8
110	~0

The rate is given in units of $10^{-9} \times$ moles/cm^{2}sec.

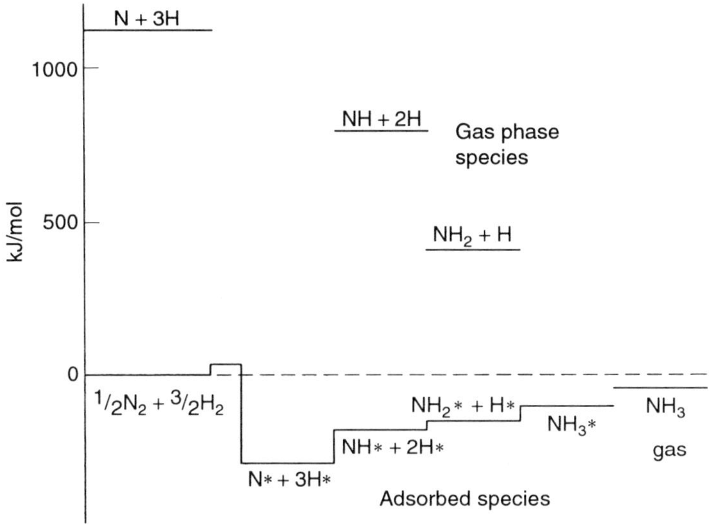

FIGURE 2.5 Energetics for the ammonia syntheses in the gas phase and at a surface.

crystal and translating the origin such that atom number 6 (see Fig. 2.6) is in (0,0,0) in the new coordinate system. Thus we can show (exercise 2.4) that the new coordinates are given by

$$\begin{pmatrix} x' \\ y' \\ z' \end{pmatrix} = \begin{pmatrix} \frac{1}{\sqrt{2}} & -\frac{1}{\sqrt{2}} & 0 \\ \frac{1}{\sqrt{6}} & \frac{1}{\sqrt{6}} & -\frac{2}{\sqrt{6}} \\ \frac{1}{\sqrt{3}} & \frac{1}{\sqrt{3}} & \frac{1}{\sqrt{3}} \end{pmatrix} \begin{pmatrix} x \\ y \\ z \end{pmatrix} - a \begin{pmatrix} 0 \\ \frac{1}{\sqrt{6}} \\ \frac{1}{\sqrt{3}} \end{pmatrix} \tag{2.7}$$

i.e., obtained by a translation and a rotation of the coordinate system.

We have in Table 2.2 given the new coordinates arising from this transformation. Introducing the following notation

$$\begin{aligned} \hat{x} &= x - \text{unit} = \frac{a}{2\sqrt{2}} \\ \hat{y} &= y - \text{unit} = \frac{a}{2\sqrt{6}} \\ \hat{z} &= z - \text{unit} = \frac{a}{\sqrt{3}} \end{aligned} \tag{2.8}$$

we get the lattice points shown in Table 2.3 for the 111 crystal. The first three layers are shown on Figure 2.7 and the 4th layer is identical to the first translated to $z = -3\hat{z} = -\sqrt{3}a$.

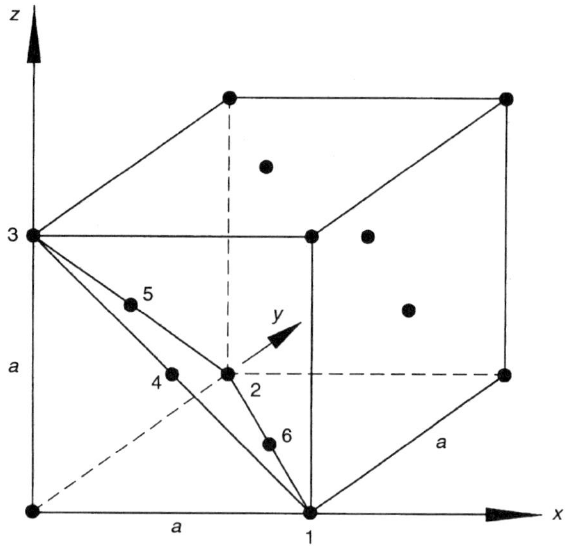

FIGURE 2.6 The (111) plane in a face-centered cubic lattice.

TABLE 2.2 Old and New Coordinates for the Pt-FCC Lattice Cut Along the 111 Plane

Point	Old Coordinate	New Coordinate
1	$a(1, 0, 0)$	$a(1/\sqrt{2}, 0, 0)$
2	$a(0, 1, 0)$	$a(-1/\sqrt{2}, 0, 0)$
3	$a(0, 0, 1)$	$a(0, -3/\sqrt{6}, 0)$
4	$a(1/2, 0, 1/2)$	$a(1/2\sqrt{2}, -3/2\sqrt{6}, 0)$
5	$a(0, 1/2, 1/2)$	$a(-1/2\sqrt{2}, -3/2\sqrt{6}, 0)$
6	$a(1/2, 1/2, 0)$	$a(0, 0, 0)$

The lattice constant $a = 3.9151$ Å [24].

TABLE 2.3 Lattice Points for the 111-Cut of a Pt-FCC Lattice

1. Layer	2. Layer	3. Layer	4. Layer
(2,0,0)	(−1,1,−1)	(−1,−1,−2)	(2,0,−3)
(−2,0,0)	(1,1,−1)	(1,−1,−2)	(−2,0,−3)
(0,−6,0)	(3,1,−1)	(0,2,−2)	(0,−6,−3)
(1,−3,0)	(−2,4,−1)	(2,2,−2)	(1,−3,−3)
(−1,−3,0)	(0,−2,−1)	(−2,2,−2)	(−1,−3,−3)
(0,0,0)	(0,4,−1)	(0,−4,−2)	(0,0,−3)

The units are given as $(\hat{x}, \hat{y}, \hat{z})$.

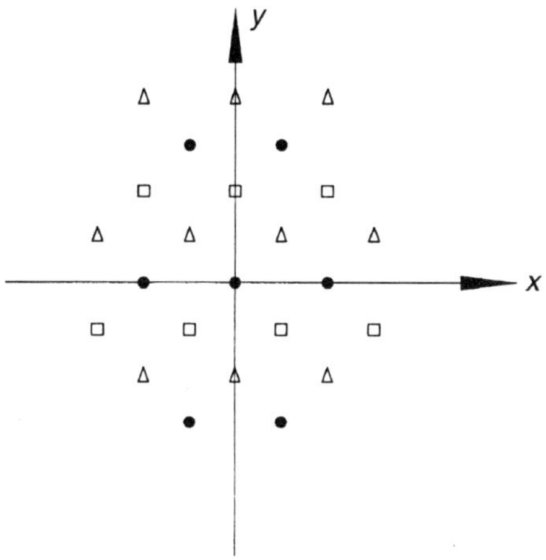

FIGURE 2.7 The position of the three top layers of the (111) cut of a face-centered cubic lattice. Top-layer atoms are indicated by black dots, second-layer atoms by triangles, and third-layer atoms by squares.

The rotation angles may in general be determined from

$$(0, 0, 1)^T = \mathbf{O}\mathbf{l} \tag{2.9}$$

where $\mathbf{l}$ is the normalized vector to be rotated. The rotation is carried out by applying the matrix $\mathbf{O}$, where $\mathbf{O}$ can be expressed by two angles α and β. The angle α specifies the amount of rotation around the original z-axis to give the coordinate system (x', y', z), β a rotation around the new x'-axis to give (x', y'', z'), i.e.,

$$\mathbf{O} = \begin{pmatrix} 1 & 0 & 0 \\ 0 & \cos\beta & \sin\beta \\ 0 & -\sin\beta & \cos\beta \end{pmatrix} \begin{pmatrix} \cos\alpha & \sin\alpha & 0 \\ -\sin\alpha & \cos\alpha & 0 \\ 0 & 0 & 1 \end{pmatrix} \tag{2.10}$$

By using the eqs. (2.9) we can determine the two angles. For simple cubic lattices the components of the vector $\mathbf{l}$ are simply given by the Miller indices. This can be used to obtain a general expression for the angles α and β (see exercise 2.8).

2.2 RECIPROCAL LATTICE SPACE

The space in which the crystals or lattices are shown in Figures 2.2 and 2.3 is called the physical or direct space. In this space the positions of the atoms can be expressed as an integral number of basis vectors $\mathbf{a}_i$ $i = 1, 2, 3$, i.e. the position of an atom (or a lattice point) is obtained by a simple translation through the vector

$$\mathbf{R}_{n_1 n_2 n_3} = n_1 \mathbf{a}_1 + n_2 \mathbf{a}_2 + n_3 \mathbf{a}_3 \tag{2.11}$$

where n_i takes all integral numbers including zero.

In a rectangular coordinate system the vectors $\mathbf{a}_i$ have the three components a_{ix}, a_{iy} and a_{iz}. Thus the vector (or lattice point) can be written as

$$\mathbf{R}_{n_1 n_2 n_3} = \mathbf{A}\mathbf{n} = \begin{pmatrix} a_{1_x} & a_{2_x} & a_{3_x} \\ a_{1_y} & a_{2_y} & a_{3_y} \\ a_{1_z} & a_{2_z} & a_{3_z} \end{pmatrix} \begin{pmatrix} n_1 \\ n_2 \\ n_3 \end{pmatrix} \tag{2.12}$$

The basis vectors are given for the three cubic lattices and the closed-packed hexagonal (CPH) in Table 2.4 (see also Fig. 2.3).

It will also be necessary to introduce the so-called *reciprocal lattice space*. Since the lattice is periodic we can represent any function as, for example, the potential, which depends upon the lattice positions, by a Fourier series or a Fourier integral,

$$V(x, y, z) = \frac{1}{(2\pi)^{3/2}} \int dk_x \int dk_y \int dk_z \overline{V}(k_x, k_y, k_z) \exp(i\mathbf{k} \cdot \mathbf{r}) \tag{2.13}$$

TABLE 2.4 Basis Vectors of Direct and Reciprocal Space for the Three Cubic Lattices: SC (Simple Cubic), BCC (Body-Centered Cubic), FCC (Face-Centered Cubic), CPH (Close-Packed Hexagonal)

Lattice	Direct **A**	Reciprocal **B**	Ω	Nearest	Next Nearest
SC	a (1,0,0)	$2\pi/a$ (1,0,0)	a^3	a (6)	$\sqrt{2}a$ (12)
	a (0,1,0)	$2\pi/a$ (0,1,0)			
	a (0,0,1)	$2\pi/a$ (0,0,1)			
BCC	$a/2$ $(-1, 1, 1)$	$2\pi/a$ (0,1,1)	$\frac{1}{2}a^3$	$\frac{\sqrt{3}}{2}a(8)$	a (6)
	$a/2$ $(1, -1, 1)$	$2\pi/a$ (1,0,1)			
	$a/2$ $(1, 1, -1)$	$2\pi/a$ (1,1,0)			
FCC	$a/2$ (0,1,1)	$2\pi/a$ $(-1, 1, 1)$	$\frac{1}{4}a^3$	$\frac{1}{\sqrt{2}}a(12)$	a (6)
	$a/2$ (1,0,1)	$2\pi/a$ $(1, -1, 1)$			
	$a/2$ (1,1,0)	$2\pi/a$ $(1, 1, -1)$			
CPH	a (1,0,0)	$\frac{2\pi}{a}(1, -\frac{1}{\sqrt{3}}, 0)$	$\frac{\sqrt{3}}{2}a^2 b$	a (6)	$\sqrt{\frac{1}{3}a^2 + \frac{1}{4}b^2}(6)$
	$\frac{a}{2}(1, \sqrt{3}, 0)$	$\frac{4\pi}{\sqrt{3}a}(0, 1, 0)$			
	$b(0,0,1)$	$\frac{2\pi}{b}(0, 0, 1)$			

The lattice constant is a and the distance between the layers in the CPH lattice is b. $\Omega = \mathbf{a}_1 \cdot \mathbf{a}_2 \times \mathbf{a}_3$ is the volume of the unit cell (see also exercise 2.1). The nearest and next-nearest neighbor distances are also given. The number in parentheses is the number of neighbors.

where $\mathbf{k} = (k_x, k_y, k_z)$ and $\mathbf{r} = (x, y, z)$. The $\mathbf{k}$ space is then the reciprocal space and the functions $\overline{V}$ can be obtained as the inverse transform

$$\overline{V}(\mathbf{k}) = \frac{1}{(2\pi)^{3/2}} \int dx \int dy \int dz\, V(x, y, z) \exp(-i\mathbf{k} \cdot \mathbf{r}) \tag{2.14}$$

The basis-vectors in the reciprocal space $\mathbf{b}_i$ can now be introduced as

$$\mathbf{b}_i = \frac{2\pi}{\mathbf{a}_i \cdot (\mathbf{a}_j \times \mathbf{a}_k)} \mathbf{a}_j \times \mathbf{a}_k \tag{2.15}$$

where i, j, k take the values (1,2,3), (2,3,1), and (3,1,2). The reciprocal vectors satisfy the relation

$$\mathbf{a}_i \cdot \mathbf{b}_j = 2\pi\delta_{ij} \tag{2.16}$$

This follows from the periodicity of the potential in direct space, i.e.,

$$V(x, y, z) = V(\mathbf{r} + n_1\mathbf{a}_1 + n_2\mathbf{a}_2 + n_3\mathbf{a}_3) \tag{2.17}$$

and a similar periodicity

$$\overline{V}(\mathbf{k}) = \overline{V}(\mathbf{k} + m_1\mathbf{b}_1 + m_2\mathbf{b}_2 + m_3\mathbf{b}_3) \tag{2.18}$$

in the reciprocal or Fourier space. The reciprocal lattice points can be expressed in a similar fashion to eq. (2.12) by introducing the B-matrix such that they are expressed as

$$\mathbf{m}^T\mathbf{B} = (m_1, m_2, m_3) \begin{pmatrix} b_{1_x} & b_{1_y} & b_{1_z} \\ b_{2_x} & b_{2_y} & b_{2_z} \\ b_{3_x} & b_{3_y} & b_{3_z} \end{pmatrix} \tag{2.19}$$

where T indicates a transposed vector (a row vector). The connection between the two sets of basis vectors can then be expressed as

$$\mathbf{BA} = 2\pi\mathbf{I} \tag{2.20}$$

where $\mathbf{I}$ is the unit matrix.

The reciprocal lattice point defined by $\mathbf{l}^T\mathbf{B}$, where the components of the vector $\mathbf{1}$ are the Miller indices, corresponds to planes in direct space specified by the Miller indices. The normal to these planes is $\mathbf{1}^T\mathbf{B}$ and hence the distance between such neighboring planes is

$$d_l = 2\pi/|\mathbf{l}^T\mathbf{B}| \tag{2.21}$$

This may be used for solving exercise 2.1.

X-ray scattering gives information on the scattering amplitude expressed as the Fourier transform of the electron density. Hence, if inversion of the data is possible, we can obtain an electron density map. Thus we experimentally obtain information about the so-called structure factor $F(k_x k_y k_z)$, where

$$F(k_x k_y k_z) = \int dx\, dy\, dz\, \rho(x, y, z) \exp(i\mathbf{k} \cdot \mathbf{r}) \tag{2.22}$$

and ρ is the electron density.

Considering elastic scattering from a collection of atoms located at $\mathbf{R}_i$ we can write the potential as

$$\sum_i V(\mathbf{x} - \mathbf{R}_i) \tag{2.23}$$

where $\mathbf{x}$ is the coordinate of the particle being scattered. The potential is "localized" in the vicinity of the atoms and small in regions between them. Introducing a Fourier transform we get

$$\sum_i V(\mathbf{x} - \mathbf{R}_i) = \sum_{i,\mathbf{k}} f_i \exp(i\mathbf{k} \cdot \mathbf{R}_i) \tag{2.24}$$

where f_i is the "scattering factor," i.e., the scattering strength of center i. For elastic scattering the initial and final energy of the scattered particle/wave is unchanged. If the mass of the scattered particle is small compared to that of the target then energy will not be deposited in the scatterer. This is of course fulfilled for electromagnetic waves but also with good approximation for light particles such as neutrons or helium atoms. If the target moves, due to lattice vibrations present at temperatures $T > 0$ K, a correction for this is needed. This correction factor is often denoted the Debye-Waller factor and will be discussed in a later section.

The wave function of the incoming particle/wave is $\exp(i\mathbf{k}_1 \cdot \mathbf{r})$ or in bra/ket notation $|\mathbf{k}_1\rangle$. The state of the scattered particle is denoted $|\mathbf{k}_2\rangle$ and for elastic scattering we have $k_1 = k_2$, i.e., only the direction of the vector is changed.

The scattering amplitude for scattering from state $|\mathbf{k}_1\rangle$ to $|\mathbf{k}_2\rangle$ is given by the matrix element

$$\left\langle \mathbf{k}_2 \middle| \sum_{i,\mathbf{k}} f_i \exp(i\mathbf{k} \cdot \mathbf{R}_i) \middle| \mathbf{k}_1 \right\rangle \tag{2.25}$$

which gives the total scattering amplitude

$$F = \sum_i f_i \exp(i\mathbf{k} \cdot \mathbf{R}_i) \tag{2.26}$$

where $\mathbf{k} = \mathbf{k}_2 - \mathbf{k}_1$. We now introduce the Laue condition, and since $\mathbf{e}_i$ are unit vectors in the direction of the incoming and outgoing directions and $\mathbf{k}_i = k$ $(i = 1, 2)$, we have $\mathbf{e}_2 - \mathbf{e}_1 = (\mathbf{k}_2 - \mathbf{k}_1)/k$. Using eq. (2.11) for $\mathbf{R}_i$ we get the following expression for F in terms of the Miller indices

$$F(h, k, l) = \sum_i f_i \exp(i2\pi(n_1 h + n_2 k + n_3 l)) \tag{2.27}$$

where we have used the fact that the wavenumber is $k = 2\pi/\lambda$.

X-ray scattering experiments measures the structure factor $F(h,k,l)$ or strictly the intensity I of diffraction spots proportional to the square of the structure factor, i.e., $I \sim |F(h, k, l)|^2$. In order to obtain f_i and thereby the atomic positions using eq. (2.24) we also need the phase of the structure factor. Various techniques have therefore been developed for solving this so-called phase problem.

The X-ray scattering explores the 3D-structure of matter but for molecule surface scattering the incoming molecule will see a surface which is periodic in the x-y plane only. Thus the atom-surface potentials are expanded in a Fourier series in 2-dimensional reciprocal lattice vectors as:

$$\begin{aligned} V(\mathbf{R},z) &= \sum_{\mathbf{G}} v_{\mathbf{G}}(z) \exp(i\mathbf{G}\cdot\mathbf{R}) = v_{00}(z) + 2v_{10}(z)\left(\cos\frac{2\pi x}{a} + \cos\frac{2\pi y}{a}\right) \\ &\quad + 2v_{11}(z)\left(\cos\frac{2\pi}{a}(x+y) + \cos\frac{2\pi}{a}(x-y)\right) + \ldots \end{aligned} \tag{2.28}$$

where $\mathbf{R} = (x, y)$, $\mathbf{G} = n\mathbf{G}_1 + m\mathbf{G}_2$, $a = b$ a lattice constant, n and m integers, and $\mathbf{G}_i$ $(i = 1, 2)$ vectors such that $\mathbf{G}_1 \cdot (x, y) = (2\pi/a)x$ and $\mathbf{G}_2 \cdot (x, y) = (2\pi/a)y$. As we shall see later, (n, m) labels the surface diffraction channels.

For studying surface structures and dynamic processes at surfaces other experimental techniques, which probe only the surface or the first few layers, are needed. Such tools are, for example, LEED (low-energy electron diffraction) or He-atom scattering (HAS) [25]. Using molecular beam technique a monoenergetic beam with an energy spread of about 0.2 meV can be produced. Time-of-flight detection technique furthermore gives insight into the distribution of energy in translational and internal degrees of freedom of the scattered particles. Thus low-energy HAS has become a very useful tool for studying various surface processes: diffraction, phonon excitation, defects, mobility of ad-atoms, and molecules [25]. Other experimental techniques which are being used for investigating surface structure are scanning tunneling microscopy (STM) and electron energy loss spectroscopy (EELS).

EXERCISES

2.1: Show that for a lattice with lattice vectors **a, b, c** and crystal angles α between **b** and **c**, β between **a** and **c**, and finally γ between vectors **a** and **b**, we have the following expressions for the volume of a unit cell

$$V = abc\sqrt{1 - \cos^2\alpha - \cos^2\beta - \cos^2\gamma + 2\cos\alpha\cos\beta\cos\gamma} \tag{2.29}$$

and for the distance d_{hkl} between planes

$$\begin{aligned}d_{hkl}^2 &= [1-\cos^2\alpha-\cos^2\beta-\cos^2\gamma+2\cos\alpha\cos\beta\cos\gamma]\\ &\quad /[(h^2/a^2)\sin^2\alpha+(k^2/b^2)\sin^2\beta+(l^2/c^2)\sin^2\gamma\\ &\quad +2(hl/ac)(\cos\alpha\cos\gamma-\cos\beta)+2(hk/ab)(\cos\alpha\cos\beta-\cos\gamma)\\ &\quad +2(kl/bc)(\cos\beta\cos\gamma-\cos\alpha)] \qquad (2.30)\end{aligned}$$

where a,b,c are lattice constants and h,k,l the Miller indices.

2.2: Calculate the De Broglie wavelength of a 20 meV He-atom.

2.3: Derive the Laue equation and from here the Bragg condition.

2.4: Find the rotated basis vectors for a (111) cut of the FCC lattice.

2.5: Calculate the reflection angles for the following planes in an FCC lattice: (111), (200), (220), and (311). The wavelength of the X-ray is 1.541 Å and the lattice constant $a = 3.62$ Å.

2.6: Calculate the rotation matrix for transforming to the (110) cut of a surface.

2.7: Determine the surface density of atoms for a simple cubic and a face-centered cubic crystal at the cuts (100), (110), and (111).

2.8: By setting $\mathbf{1} = (h, k, l)/\sqrt{h^2+k^2+l^2}$, derive a general expression for the angles α and β.

2.3 LATTICE DYNAMICS

The solution of the dynamical problem for the gas and surface atoms requires in principle solution of the quantum mechanical equations of motion for the system. Since this problem has been solved only for 3–4 atomic systems we need to incorporate some approximations. One obvious suggestion is to treat the dynamics of the heavy solid atoms by classical rather than quantum dynamical equations. As far as the lattice is concerned we may furthermore take advantage of the periodicity of the atom positions. At the surface this periodicity is, however, broken in one direction and special techniques for handling this situation are needed. Lattice dynamics deals with the solution of the equations of motion for the atoms in the crystal. As a simple example we consider first a one-dimensional "crystal" of atoms with identical masses. If we include only the nearest neighbor interaction, the hamiltonian is given by:

$$H = \sum_{j=1}^{N}\frac{m}{2}\dot{\xi}_j^2 + \frac{1}{2}\sum_{j=2}^{N}F(\xi_j-\xi_{j-1})^2 \qquad (2.31)$$

where F is the force constant, m the mass of the atoms, ξ_j the displacement coordinate for atom number j, and $\dot{\xi}_j = (d\xi_j)/(dt)$ the velocity. By applying

Newton's second law, we get the equations of motion for the atom-displacements as

$$m\frac{d^2\xi_j}{dt^2} = F(\xi_{j+1} + \xi_{j-1} - 2\xi_j) \tag{2.32}$$

for $j = 2, \ldots N - 1$. For $j = 1$ and N we have

$$m\frac{d^2\xi_1}{dt^2} = F(\xi_2 - \xi_1) \tag{2.33}$$

$$m\frac{d^2\xi_N}{dt^2} = F(\xi_{N-1} - \xi_N) \tag{2.34}$$

Introducing now $\xi_j(t) = \exp(i(\omega t + j\phi))$ and the wavenumber $k = 2\pi/\lambda$ such that

$$\phi = \frac{2\pi a}{\lambda} = ka \tag{2.35}$$

we get the so-called dispersion relation, relating the frequency to the wavenumber

$$\omega(k) = \pm\sqrt{\frac{4F}{m}}\sin\left(\frac{1}{2}ka\right) \tag{2.36}$$

where a is the distance between the atoms. This expression is often referred to as the dispersion relation for Rayleigh surface waves [92]. The dispersion of the Rayleigh wave can be measured by, for example, electron energy loss spectroscopy [93]. We notice that for small values of k we obtain the simple dispersion law $\omega(k) = \sqrt{F/m}ka$. The group velocity of the waves with long wavelength ($k \to 0$) is then

$$v_g = \frac{d\omega(k)}{dk} = a\sqrt{F/m} \tag{2.37}$$

The dispersion curve (Fig. 2.8) corresponds to the so-called low frequency acoustical mode. In a three-dimensional lattice this mode survives and appears as a surface wave.

If the masses in the one-dimensional lattice are different with alternating values m_1 and m_2, one obtains by solving the lattice dynamical equations the following dispersion relation:

$$\omega = \pm\sqrt{F/m}\left[1 \pm \left(1 - \frac{4m_1 m_2 \sin^2 ka}{(m_1 + m_2)^2}\right)^{1/2}\right]^{1/2} \tag{2.38}$$

where $m = m_1 m_2/(m_1 + m_2)$. The high-frequency branch is called the optical branch. If the two masses belong to ions as, for example, in the NaCl crystal, the vibrational motion in the optical mode produces an oscillating dipole, which is infrared active, i.e, it absorbs light in the infrared.

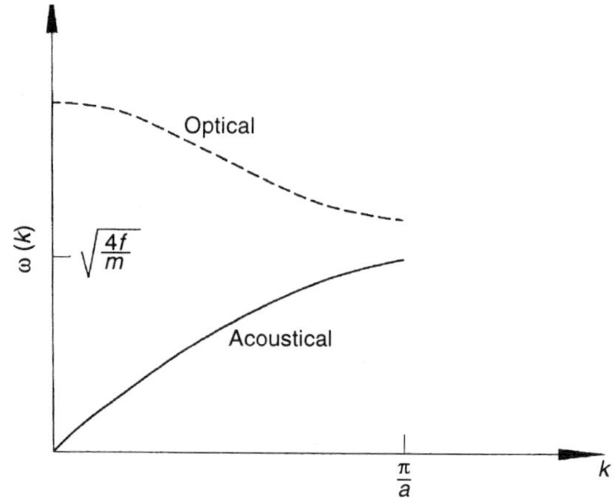

FIGURE 2.8 The dispersion curves for a simple one-dimensional lattice with an optical and an acoustical branch.

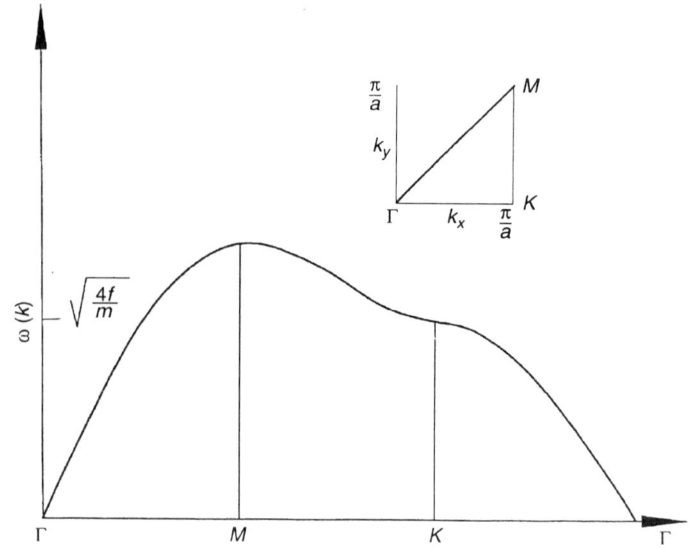

FIGURE 2.9 The dispersion curve in the directions of the reciprocal lattice vectors for a two-dimensional lattice.

Also for a two-dimensional lattice with lattice constant $a = b$ one may solve the lattice dynamical equations analytically and obtain the dispersion relation

$$\omega = \pm\sqrt{\frac{2F}{m}}(2 - \cos k_x a - \cos k_y a)^{1/2} \tag{2.39}$$

The dispersion relation for this doubly degenerate transverse acoustical mode is also shown in Figure 2.9. The relation (2.39) indicates that the dispersion curve depends upon the direction in the reciprocal lattice space (k_x, k_y). Only for the simplistic models such as those mentioned above is it possible to obtain analytical expressions for the dispersion curves. For the general three-dimensional situation the lattice dynamics must be solved numerically.

The linear chain model has also been used to study the role of energy dissipation in atom surface collisions. It is of course too simplistic to approximate the surface by a linear chain of atoms but the model is easily solvable and can give some qualitative insight into the problem. Thus the problem can be formulated by adding the following equation to the set of equations for the spring or chain atoms

$$M\frac{d^2z}{dt^2} = -\frac{\partial}{\partial z}V(z - R_1) \tag{2.40}$$

where R_1 is the position of atom 1 (the end atom) in the chain and $V(x)$ a potential depending on the distance between the incoming atom with mass M and the outermost chain atom. In addition, the equation for atom 1 is modified to

$$m\frac{d^2R_1}{dt^2} = F(R_1 - R_2) - \frac{\partial}{\partial R_1}V(z - R_1) \tag{2.41}$$

and for the remaining atoms $j = 2, \ldots$ we have

$$m\frac{d^2R_j}{dt^2} = F(R_{j+1} + R_{j-1} - 2R_j) \tag{2.42}$$

These equations have been considered by Zwanzig [21]. He reduced the solution of the many-body problem to one equation

$$\frac{d^2u(\tau)}{d\tau^2} = \frac{1+\mu}{4\mu}f(u(\tau)) - \frac{1}{2}\int_0^\tau ds\frac{J_2(s)}{s}f(u(\tau - s)) \tag{2.43}$$

where J_2 is a Bessel function, $\mu = M/m$ is the ratio between the gas atom and the surface atom. τ is a reduced time $\tau = 2\omega t$ where $\omega = \sqrt{F/m}$. The relative distance is $u(\tau) = z(t) - R_1(t)$ and the reduced force

$$f(u) = -\frac{1}{F}\frac{\partial V(z - R_1)}{\partial z} \tag{2.44}$$

We notice that the integral in eq. (2.43) involves some memory through the integration over the force from time zero to time τ. Zwanzig solved the above equation for some simple potentials. For more general potentials as, for instance, an exp-6 potential with a damping of the long-range term (see also Chapter 4), i.e.,

$$V(z) = C\exp(-az) - C_6z^{-6}(1 - \exp(-bz))\sum_{k=0}^{6}\frac{(bz)^k}{k!} \tag{2.45}$$

the equation has to be integrated numerically. In Table 2.5 we have shown some results obtained with masses and force constants corresponding to the collision of He or Ne atoms with a chain of copper atoms. We notice that energy transfer E_{int} to the "crystal" is not negligible despite the large mass ratio between the helium and the copper atom. For the heavier neon atom the energy transfer increases and amounts to as much as 37% at 40 kJ/mol. This is in agreement with findings using the more realistic three-dimensional lattices and, for instance, classical trajectory calculations (see also Chapters 6 and 8). For a collision on a real surface, the atom may then be trapped after a few collisions.

2.3.1 Lattice Dynamics in 3D-Crystals

We now consider a convenient formulation of the dispersion problem in the case of a bulk crystal and in the presence of a surface. The position of an atom k in a given unit cell l can be expressed as

$$\mathbf{x}(lk) = \mathbf{x}_0(lk) + \xi(lk) \tag{2.46}$$

where ξ is the displacement from equilibrium position $\mathbf{x}_0$. The position of the unit cell $\mathbf{l}$ may be written as:

$$\mathbf{x}(\mathbf{l}) = \sum_i l_i \mathbf{a}_i \tag{2.47}$$

where $\mathbf{a}_i$ with $i = 1,2,3$ are the lattice unit vectors and $\mathbf{l} = (l_1, l_2, l_3)$. The total kinetic energy of the lattice motion is given by

$$T = \frac{1}{2} \sum_{lk\alpha} M_k \dot{\xi}_\alpha^2(lk) \tag{2.48}$$

TABLE 2.5 Energy Transfer E_{int} to a Linear Chain of Atoms with Mass 63.5 amu for Collisions with an Atom with Mass 4 amu and 20.2 amu

Initial Kinetic Energy	Final Kinetic Energy	E_{int}	α	Atom
10.0	8.93	1.07	0.11	He
20.0	17.28	2.72	0.14	He
40.0	33.45	6.55	0.16	He
10.0	7.78	2.22	0.22	Ne
20.0	14.17	5.83	0.29	Ne
40.0	25.17	14.83	0.37	Ne

The interaction potential is a damped exp-6 potential with parameters from ref. [40]. The force constant F was obtained from the Debye frequency corresponding to a Debye temperature 315 K. Energies in kJ/mol. α is the energy accomodation coefficient E_{int}/E_{kin}, where E_{kin} is the initial kinetic energy.

where α labels the cartesian coordinate and M_k is the mass of atom k. The potential energy is expanded (usually just to second order) around the equilibrium position defined by

$$V_\alpha(lk) = \left.\frac{\partial V}{\partial \xi_\alpha(lk)}\right|_0 = 0 \tag{2.49}$$

Thus we have

$$V = V_0 + \frac{1}{2} \sum_{lk\alpha;l'k'\beta} V_{\alpha\beta}(lk;l'k')\xi_\alpha(lk)\xi_\beta(l'k') + \ldots \tag{2.50}$$

where $V_{\alpha\beta}$ is the second derivative of the potential evaluated at the equilibrium positions. In the harmonic approximation the higher-order terms in eq. (2.50) are neglected. Not much is known about the influence of anharmonic terms on molecule surface scattering, but phonon-phonon coupling arising from such terms is without doubt important for the dissipation of energy in the crystal and for properties depending on surface temperature. For a more complete discussion of anharmonic corrections, the reader is referred to ref. [22].

The equations of motion are now obtained as

$$M_k \frac{d^2\xi_\alpha(lk)}{dt^2} = -\frac{\partial V}{\partial \xi_\alpha(lk)} = -\sum_{l'k'\beta} V_{\alpha\beta}(lk;l'k')\xi_\beta(l'k') \tag{2.51}$$

In order to solve this equation we assume the following form for the displacement of atom k in unit cell $\mathbf{l}$ in directions $\alpha = x, y, z$

$$\xi_\alpha(\mathbf{l}k) = M_k^{-\frac{1}{2}} \xi_\alpha(k) \exp(i(\mathbf{K} \cdot \mathbf{x}(\mathbf{l}) - \omega t)) \tag{2.52}$$

where $\mathbf{K} = (K_x, K_y, K_z)$ and $\mathbf{x}(\mathbf{l})$ is defined by eq. (2.47). If we insert this expression in eq. (2.51) we get:

$$\omega^2 \xi_\alpha(k) = \sum_{k'\beta} A_{\alpha\beta}(kk', \mathbf{K})\xi_\beta(k') \tag{2.53}$$

where the elements of the matrix A are given by

$$A_{\alpha\beta}(kk', \mathbf{K}) = \frac{1}{\sqrt{M_k M_{k'}}} \sum_{\mathbf{L}'} V_{\alpha\beta}(lk;l'k') \exp(-i(\mathbf{x}(\mathbf{l}) - \mathbf{x}(\mathbf{l}')) \cdot \mathbf{K}) \tag{2.54}$$

It should be noted that the matrix $\mathbf{A}$ is independent of $\mathbf{l}$ (for an infinite system) since:

$$V_{\alpha\beta}(\mathbf{l}k;\mathbf{l}'k') = V_{\alpha\beta}(0k;\mathbf{l}' - \mathbf{l}, k') \tag{2.55}$$

due to the translational invariance from unit cell to unit cell. Equation (2.53) may be written in matrix notation

$$\omega^2\xi = \mathbf{A}\xi \tag{2.56}$$

where ω^2 is a diagonal matrix

$$[\omega^2]_{\alpha k;\beta k'} = \omega^2\delta_{\alpha\beta}\delta_{kk'} \tag{2.57}$$

The eigenvalue equation

$$|A_{pq} - \omega^2\delta_{pq}| = 0 \tag{2.58}$$

is the condition for nontrivial solutions to (2.56) and determines the dispersion curves in a 3D lattice. The eqs. (2.58) are of the dimension $3r$ in ω^2, where r is the number of atoms in the unit cell.

In the presence of a *surface*, we slightly modify the above scheme. Thus the vectors $\mathbf{a}_1$ and $\mathbf{a}_2$ are taken to be parallel to the surface. The vector $\mathbf{a}_3$ spans the dimension perpendicular to the surface. The position of atom k may be written as

$$\mathbf{X}_0(lk) = \mathbf{X}(\mathbf{L}) + \mathbf{X}(l_3k) \tag{2.59}$$

where

$$\mathbf{X}(\mathbf{L}) = l_1\mathbf{a}_1 + l_2\mathbf{a}_2 \quad \mathbf{L} = (l_1, l_2) \tag{2.60}$$

and $\mathbf{X}(l_3k) = l_3\mathbf{a}_3$. Here l_3 denotes the layer parallel to the surface $l_3 = 1, 2, \ldots N_L$. Strictly speaking we should have used the notation $l_i(k)$, but this has been dropped for the sake of simplicity.

The equations of motion now read

$$M_k(l_3)\frac{d^2\xi_\alpha(\mathbf{L}, l_3k)}{dt^2} = -\sum_{L'l_3'k'\beta} V_{\alpha\beta}(\mathbf{L} - \mathbf{L}', l_3kl_3'k')\xi_\beta(\mathbf{L}', l_3', k') \tag{2.61}$$

where we have used the fact that there is a periodicity in the planes labeled by l_3.

Possible solutions to eq. (2.61) are plane waves traveling along the surface, i.e.,

$$\xi(\mathbf{X}, l_3k) = \sqrt{\frac{\hbar}{2N_s\omega_j(\mathbf{K})M_k(l_3)}}\,\mathbf{e}(l_3k \mid \mathbf{K}j)\exp(i(\mathbf{K}\cdot\mathbf{X}(\mathbf{L}) - \omega_j(\mathbf{K})t)) \tag{2.62}$$

where N_s is the number of unit cells, i.e., the number of values taken by the vector index $\mathbf{L}$, and $\mathbf{e}(l_3k|\mathbf{Kj})$, the polarization vector, is an eigenvector to the secular equation

$$\sum_{l_3'k'\beta} A_{\alpha\beta}(l_3k; l_3'k'; \mathbf{K})e_\beta(l_3'k'|\mathbf{K}j) = \omega_j^2(\mathbf{K})e_\alpha(l_3k|\mathbf{K}j) \tag{2.63}$$

where

$$A_{\alpha\beta} = [M_k(l_3)M_{k'}(l'_3)]^{-1/2} \sum_{\mathbf{L}} V_{\alpha\beta}(\mathbf{L} - \mathbf{L}', l_3k, l'_3k') \exp(i\mathbf{K} \cdot [\mathbf{X}(\mathbf{L}) - \mathbf{X}(\mathbf{L}')]) \tag{2.64}$$

For this so-called slab solution to the dynamical problem we have to solve a secular equation of the order $(3rN_L)^2$ and we obtain therefore $3rN_L$ values of the frequency $\omega_j(\mathbf{K})$ for each value of **K**. Dispersion curves have been obtained as a function of the **K** direction in the reciprocal lattice space, using the slab method by, for instance, Allen et al. [23]. It should be noted that a different numerical method for obtaining the dispersion curves using Green's functions has been introduced by Benedek [26]. The calculated dispersion curves can be compared with those obtained experimentally using, for example, HREELS (high-resolution electron energy loss spectroscopy); see, for example, [27].

2.3.2 Phonons (Normal Mode Solution)

In order to describe lattice motion it is convenient to introduce phonons. The theoretical discussion of phonons dates back to the papers by Lord Rayleigh, who considered waves propagating along the surface of an elastic continuum [92]. These modes are also found among the low-frequency modes obtained when solving the normal mode problem of a solid. Within the harmonic approximation the hamiltonian for the solid is

$$H = V_0 + \frac{1}{2} \sum_{lk\alpha} M_k \dot{\xi}_\alpha^2(lk) + \frac{1}{2} \sum_{lk\alpha;l'k'\beta} V_{\alpha\beta}(lk; l'k')\xi_\alpha(lk)\xi_\beta(l'k') \tag{2.65}$$

where M_k is the atom mass and $\xi_\alpha(kl)$ the displacement from the equilibrium position of atom k in unit cell (l) and direction α. We may introduce some new coordinates $Q(\mathbf{K}j)$ such that

$$\xi_\alpha(lk) = (NM_k)^{-1/2} \sum_{\mathbf{K},j} e_\alpha(k; \mathbf{K}j) Q(\mathbf{K}j) \exp(i\mathbf{K} \cdot \mathbf{X}(l)) \tag{2.66}$$

where $Q(\mathbf{K}j)$ are the normal mode coordinates and $e_\alpha(k; \mathbf{K}j)$ elements of the eigenvectors to eq. (2.56). Inserting (2.66) in the hamiltonian (2.65), we get the kinetic energy as

$$T = \frac{1}{2} \sum_{\mathbf{K}j} \dot{Q}^T(\mathbf{K}j)\dot{Q}(\mathbf{K}j) \tag{2.67}$$

and the potential energy

$$V = V_0 + \frac{1}{2} \sum_{\mathbf{K}j} \omega_j^2 Q^T(\mathbf{K}j) Q(\mathbf{K}j) \tag{2.68}$$

where T denote the transposed vector. Thus the hamiltonian is diagonal and separable in the normal modes, and the $3rN$ equations of motion are

$$\ddot{Q}(\mathbf{K}j) + \omega_j^2(\mathbf{K})Q(\mathbf{K}j) = 0 \tag{2.69}$$

Here r is the number of particles in a unit cell and N the number of unit cells in the crystal. The phonons follow Bose-Einstein statistics, and hence the thermal population distribution at the temperature T is given by

$$n(\mathbf{K}j; T) = [\exp(\hbar\omega_j(\mathbf{K})/k_B T) - 1]^{-1} \tag{2.70}$$

With the above distribution, the average energy of a given mode can be obtained as the Boltzmann weighted average over the harmonic oscillator energies

$$\varepsilon = \sum_{n=0}^{\infty} \hbar\omega \left(n + \frac{1}{2}\right) \exp\left(-\frac{\hbar\omega}{k_B T}(n + 1/2)\right) \tag{2.71}$$

where we have simplified the ω notation. The average energy can also be expressed as

$$\varepsilon = k_B T^2 \frac{\partial z}{\partial T} \tag{2.72}$$

where z is the partition function defined by

$$z = \sum_{n=0}^{\infty} \exp(-\hbar\omega(n + 1/2)/k_B T) = \frac{1}{2\sinh(\eta/2)} \tag{2.73}$$

where $\eta = \hbar\omega/k_B T$. Using this expression we obtain

$$\varepsilon(\omega) = \frac{\hbar\omega}{2} + \frac{\hbar\omega}{\exp\eta - 1} \tag{2.74}$$

We notice that for $T \to 0$ we have $\varepsilon \to \hbar\omega/2$ (the zero point energy) and for $T \to \infty$ we have $\varepsilon \to k_B T$. This limit is a manifestation of the so-called equipartition theorem according to which the ensemble averaged energy in each degree of freedom in phase space (i.e., both momentum and coordinate space) is $(1/2)k_B T$.

We shall return to a dynamical treatment of phonon dynamics when dealing with atom/molecule surface scattering.

The surface-localized vibrations are low-frequency acoustical modes (Rayleigh modes) which are easily excited by collisions with atoms and molecules. The bulk modes are often approximated by the Debye model where the distribution is given as [98]

$$g(\nu) = 9N\frac{\nu^2}{\nu_D^3} \tag{2.75}$$

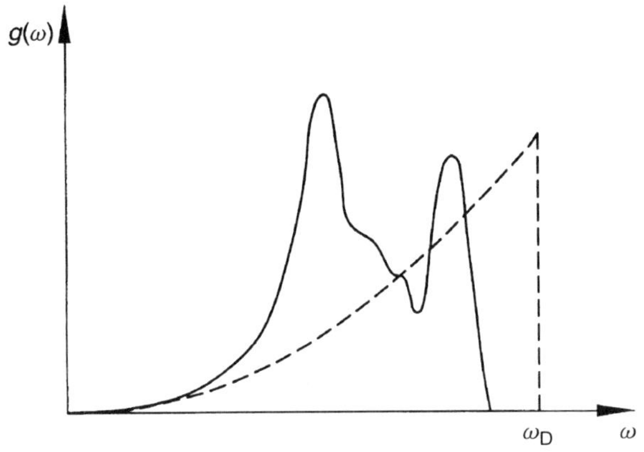

FIGURE 2.10 The phonon distribution $g(\omega)$ for the Debye model (dashed line). An experimentally obtained curve is also shown (solid line).

where $\nu = \omega/2\pi$ and N is the number of atoms, such that

$$\int_0^{\nu_D} g(\nu)\, d\nu = 3N \tag{2.76}$$

$3N$ being the number of vibrational modes. Thus the normalized to unity distribution is $3\nu^2/\nu_D^3$. Historically, the Debye model [115] was introduced to explain the low temperature heat capacity, which was incorrect in the Einstein model (see Fig. 2.10), where all the atoms in the crystal were assumed to vibrate with the same frequency.

According to the Debye model the low temperature behavior of the heat capacity is $\sim T^3$. However, due to a contribution from the electrons we have the following low temperature behavior

$$C_V \sim aT + bT^3 \tag{2.77}$$

where $a = \pi^2 N k_B / 2T_F$ for a free electron gas. Here k_B is Boltzmann's constant and T_F the Fermi temperature. The Fermi temperature lies in the range 10^4 to 10^5 K and the Debye constant b is given by

$$b = \frac{12\pi^4}{5} N k_B (1/\Theta_D)^3 \tag{2.78}$$

where N is the number of atoms in the solid. Table 2.6 gives the Debye temperatures, where

$$\Theta_D = h\nu_D/k_B \tag{2.79}$$

for some metals. Thus typical values for metals are a few hundred Kelvin.

TABLE 2.6 Debye Temperatures for Some Solids

Solid	Θ_D/K	Solid	Θ_D/K	Solid	Θ_D/K
Na	150	K	100	Mg	290
Cu	315	Ni	375	Zn	250
Ag	215	Au	170	Pb	88
Fe	420	Pt	225	Al	390

The average energy of a crystal with N atoms, following the Debye distribution, can now be obtained by using eq. (2.74)

$$\langle E \rangle = 9N \int_0^{\omega_D} \varepsilon(\omega) \frac{\omega^2}{\omega_D^3} d\omega \tag{2.80}$$

where we have introduced $\omega = 2\pi\nu$. The integral over frequencies can be carried out analytically for $\Theta_D/T \gg 1$, such that

$$\langle E \rangle = \frac{9}{8} N k_B \Theta_D + 9 N k_B T \frac{\pi^4}{15} \left(\frac{T}{\Theta_D} \right)^3 \tag{2.81}$$

and hence we obtain by using the fact that the heat capacity is given as $C_V = (\partial \langle E \rangle)/(\partial T)$, the abovementioned value for b.

The Debye model is able qualitatively to reproduce the density of the low-frequency modes, but often molecule surface dynamics are carried out using a finite-size crystal with from a few hundred to about a thousand atoms, and for this size of "crystals" the phonon spectrum can be obtained by diagonalizing the force constant matrix. This is furthermore realistic if the reactivity of "nanostructures" is to be investigated. For crystals of "macroscopic" size we can as indicated above use the periodicity of the crystal to reduce the normal mode problem to one involving just a unit cell. However, for surface scattering, the surface modes are of importance, and also the possibility of having ad-atoms, steps, and other surface anomalities play an important dynamical role. Hence the modeling using a finite-size-type crystal may often more realistically reproduce the important excitation modes for surface scattering dynamics. This approach makes it possible to study, within the same framework, energy transfer to and reactivity on polyatomic molecules, zeolite structures, and "biochemical environments" within the same computational scheme.

EXERCISES

2.9: Derive the expression (2.39) for the two-dimensional lattice.

2.10: Calculate the number of frequencies $g(\omega)d\omega$ in the range ω to $\omega + d\omega$ for the linear chain model. Use the fact that the number of frequencies is N and that $\int (1/\sqrt{a - cx^2})\, dx = (1/\sqrt{c}) \arcsin (x\sqrt{c/a})$.

2.11: Compare the contribution to C_V from electrons and phonons at 10 K for copper. The Fermi temperature is 8.2 10^4 K. At higher temperatures the following expression is obtained for the heat capacity within the Debye model

$$C_V = 9Nk_B \frac{T^3}{\Theta_D^3} \int_0^{\theta_D/T} \frac{x^4 \exp(x)}{(\exp(x) - 1)^2} dx \tag{2.82}$$

In the low temperature limit the integral becomes $4\pi^4/15$.

2.12: Estimate the force constant for a Pt-crystal from the following information on the volume change ΔV when stretching Pt-thread [116]

$$\frac{\Delta V}{V_0} = 3.110^{-7} p \tag{2.83}$$

where V_0 is the volume before applying the load p (in kg/cm^2). *Hint*: Use a simple cubic lattice and the gravity acceleration $g = 980.66$ cm/sec^2 and a lattice constant $a = 3.9151$ Å.

2.4 ATOM-SURFACE SCATTERING, KINEMATICS

In order to describe atom-surface scattering in the low energy limit where only surface waves and one-phonon excitations are important, it is convenient first to consider a simple *kinematic* approach. Figure 2.11 shows an incoming He-atom with incident wavevector $\mathbf{k}_i$. The incident polar angle is θ_i and the azimuthal angle ϕ_i. The initial energy of the atom is then given as

$$E_i = \frac{\hbar^2 k_i^2}{2m} \tag{2.84}$$

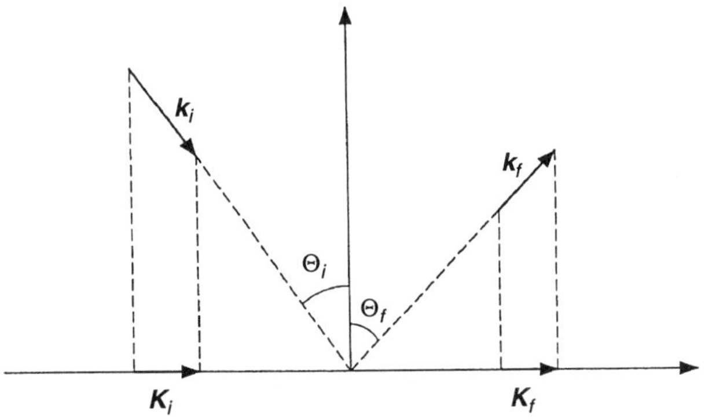

FIGURE 2.11 An atom hits the surface with the incident angle Θ_i to the surface normal and leaves with the scattering angle Θ_f. The surface-projected wavevectors are $\mathbf{K}_i$ and $\mathbf{K}_f$.

where m is the mass of the atom. After being scattered, the atom leaves the surface with the wavevector $\mathbf{k}_f$. The surface-projected wavevectors are

$$\mathbf{K}_i = \mathbf{k}_i \sin\theta_i \tag{2.85}$$

$$\mathbf{K}_f = \mathbf{k}_f \sin\theta_f \tag{2.86}$$

By varying the values of the incident energy E_i and angle θ_i, one can cover the full kinematic space according to the following two conservation rules:

Energy Conservation

$$\frac{\hbar^2 k_f^2}{2m} = \frac{\hbar^2 k_i^2}{2m} + \hbar\omega \tag{2.87}$$

where we assume a single phonon creation $\hbar\omega < 0$ and annihilation $\hbar\omega > 0$ process. Thus multiphonon excitation will be neglected for the time being.

Crystal Momentum Conservation The momentum in the parallel plane is conserved to within an reciprocal lattice vector $\mathbf{G}_{nm}$, i.e.,

$$\mathbf{K}_f = \mathbf{K}_i + \mathbf{G}_{nm} + \mathbf{Q} \tag{2.88}$$

where $\mathbf{Q}$ is the two-dimensional phonon momentum vector and the reciprocal lattice vector is

$$\mathbf{G}_{nm} = \frac{2\pi}{a}(n\mathbf{G}_1 + m\mathbf{G}_2) \tag{2.89}$$

where $\mathbf{G}_i$ are reciprocal lattice unit vectors, and a the lattice constant. Introducing $\Delta\mathbf{K} = \mathbf{Q} + \mathbf{G}_{nm}$ we get

$$k_f^2 = k_i^2 + \frac{2m\omega}{\hbar} \tag{2.90}$$

and

$$\mathbf{K}_f = \mathbf{K}_i + \Delta\mathbf{K} \tag{2.91}$$

where $\Delta\mathbf{K}$ is the total exchange of parallel momentum with the surface. Elastic scattering is defined by $\omega = \mathbf{Q} = 0$ and the indices n, m label the various diffraction channels.

Considering an experiment (Fig. 2.12) where the angle to the detector θ_d is fixed and where θ_i is varied by rotating the crystal we have

$$\theta_i + \theta_f = \theta_d \text{ (fixed)} \tag{2.92}$$

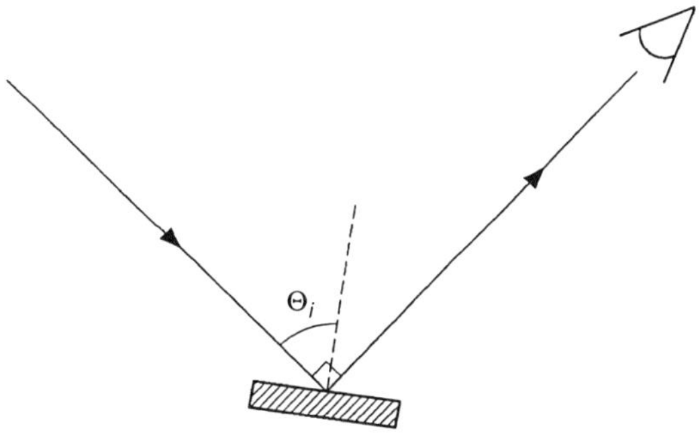

FIGURE 2.12 The detection of a scattered wave with a fixed detection angle $\theta_d = 90°$.

This relation is valid only for "in-plane" scattering. Out-of-plane scattering is small if the beam is oriented in directions of high symmetry.

The above relation gives the following connection between θ_i and θ_d:

$$\sin\theta_i = \frac{1}{2}\left[-\frac{G}{k_i} + \sqrt{\frac{2\sin^2\theta_d + G^2(\cos\theta_d - 1)/k_i^2}{1 + \cos\theta_d}}\right] \tag{2.93}$$

For $\theta_d = 90°$, this reduces to

$$\sin\theta_i = \frac{1}{2}\left[-\frac{G}{k_i} + \sqrt{2 - \frac{G^2}{k_i^2}}\right] \tag{2.94}$$

Eq. (2.94) gives a relation for the angles where the various diffraction peaks are to be expected as a function of the magnitude of the reciprocal lattice vector G. In the case of single phonon excitation we may also solve eqs. (2.87, 2.88), which in the case of in-plane scattering and $\theta_d = 90°$, gives the following constraint on the phonon frequency

$$\omega(\Delta K) = \frac{E_i}{\hbar}\left[\frac{(\sin\theta_i + \Delta K/k_i)^2}{\cos^2\theta_i} - 1\right] \tag{2.95}$$

where $E_i = \hbar^2 k_i^2/2m$.

Inelastic processes are now possible whenever the curve (2.95) $\omega(\Delta K)$ crosses a phonon dispersion curve (see Fig. 2.13). Such plots are called zone diagrams and are essential for the correct interpretation of time-of-flight (TOF) spectra of atom-surface inelastic scattering processes [117].

2.4.1 Cross Sections

In scattering theory the so-called cross section plays a central role. The cross section is associated with the probability for a certain event of the collision as

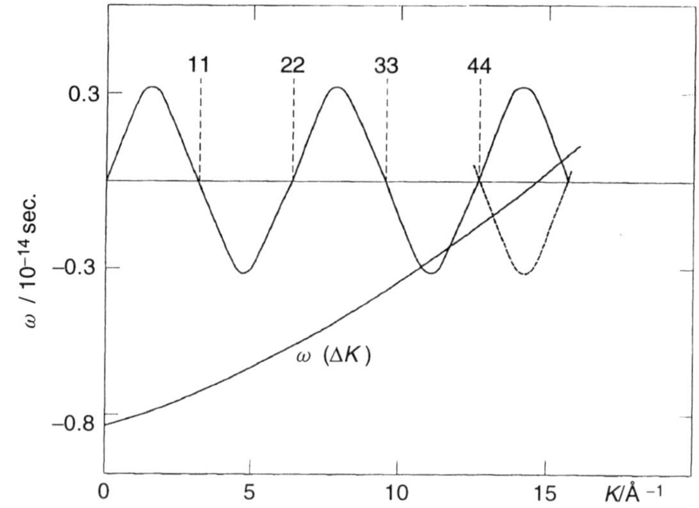

FIGURE 2.13 Dispersion curve for a 2D-lattice with $\omega = \pm\sqrt{(2f/m)}\sqrt{2 - 2\cos Ka}$, where f is the force constant, m the mass of the lattice atom, K the wavenumber, and a the distance between the atoms (upper curve). $\omega(\Delta K)$ is the curve given by eq. (2.95), where $K = G + Q$, and $G = |\mathbf{G}_{nn}|$ and $Q = |\mathbf{Q}|$ the projection of the phonon wavevector on the surface plane.

for instance a reaction, inelastic transition, or elastic transition. In the latter case, the velocity vector only changes its direction. In gas-phase collisions this velocity vector is related to the relative motion, but due to the infinite mass of the solid, this motion in surface scattering is simply the center of mass motion of the incoming molecule. The cross section depends on the initial and final values of the velocity vector, i.e., we have $\sigma(\mathbf{k}, \mathbf{k}')$, where $\mathbf{k} = m\mathbf{v}/\hbar$. Usually the notation is simplified to $\sigma(E, \theta, \phi)$ by keeping the incident angles fixed and noting that energy conservation relates k' and k, where E is the initial kinetic energy, and θ and ϕ the final (scattering) angles. The cross section has the unit of Å^2 and denotes the size of the target. More specifically, the reaction cross section would be the size of the target seen by the collider provided that the result of the collision is a reaction. Similarly we can relate it to any other specific event, elastic, inelastic, dissociation, and so forth.

In surface scattering the size of the target is in principle infinite. Hence, in order for the cross section to be specified, it has to be connected with a certain area of the surface, say an area defined by the unit cell. The quantity which can be measured is the number of particles scattered in a given solid angle and energy range.

According to scattering theory [28] the amplitude for scattering of a structureless particle from momentum state $\mathbf{k}'$ to $\mathbf{k}$ is

$$f(\mathbf{k}, \mathbf{k}') = -\frac{m}{2\pi\hbar^2}\langle \mathbf{k}'|V|\mathbf{k}^+\rangle \tag{2.96}$$

where m is the particle mass, V the interaction potential, and $|\mathbf{k}^+\rangle$ denotes the scattered state. The scattered state can be written as a series, i.e.,

$$|\mathbf{k}^+\rangle = |\mathbf{k}\rangle + \text{ higher-order terms} \tag{2.97}$$

The so-called first-order Born approximation is obtained by including just the first term in this series.

Often we are interested in the scattered flux into a certain solid angle $d\Omega = \sin\theta d\theta\, d\phi$, where the angles θ and ϕ give the direction of $\mathbf{k}$ and we can define the differential cross section as

$$d\sigma(k) = |f(\mathbf{k}, \mathbf{k}')|^2\, d\Omega \tag{2.98}$$

i.e., the amount of probability or squared amplitude in a given direction. In the Born approximation we then have

$$\frac{d\sigma}{d\Omega} = \frac{m^2}{4\pi^2\hbar^4}|\langle\mathbf{k}'|V|\mathbf{k}\rangle|^2\frac{k}{k'} \tag{2.99}$$

where the last factor is a flux factor, i.e., outgoing divided by incoming flux.

Often we also define a differential cross section not only for scattering in a certain solid angle but also in a certain energy range $E; E + dE$. This can be introduced by noting that $d\mathbf{k} = k^2dkd\Omega = (mk)/\hbar^2 dE\, d\Omega$, where we have used that $E = \hbar^2k^2/2m$. Again in the simple case considered here we define the differential cross section for scattering in a given solid angle θ, ϕ and energy range as

$$\frac{d^2\sigma}{d\Omega dE} = \frac{2m}{\hbar^2}\frac{m^2}{(2\pi)^2\hbar^4}|\langle\mathbf{k}'|V|\mathbf{k}\rangle|^2\frac{k}{k'}\delta(k^2 - 2mE/\hbar^2) \tag{2.100}$$

where $|\mathbf{k}\rangle$ is a plane wave state $\exp(i\mathbf{k}\cdot\mathbf{R})$. We notice that the delta-function has the unit of Å^2.

Scattering of helium atoms from surfaces or adsorbates combined with time-of-flight measurements has been used by Toennies and coworkers [118] to obtain information on the spectroscopy of adsorbed molecules, diffusion constants and barriers, phonon excitations, and the influence of surface temperature.

EXERCISES

2.13: Verify eq. (2.95).

2.14: Use momentum and energy conservation to obtain an approximate expression for the energy transfer to a solid

$$\Delta E = \frac{4m}{M_s}(1 - m/M_s)\,(E^g_{kin}(1 - m/M_s) - k_BT_s) \tag{2.101}$$

where m is the mass of the atom/molecule, M_s of a surface atom, E^g_{kin} the kinetic energy of the gas atom/molecule, and T_s the surface temperature. Compare the energy accommodation coefficient $\alpha = \Delta E / E^g_{kin}$ at $T_s = 0$ with the calculated values in Table 2.5 and with the Baule formula (1.27). How does increasing the surface temperature affect the value of α ?

3

SURFACE DIFFUSION

The diffusion of adsorbates on the surface is a process which is important for reactions of the Langmuir-Hinshelwood type, where the molecules in question travel some distance on the surface before they react. The surface diffusion process is influenced by the structure and corrugation of the surface. Therefore the information which can be obtained by studying the diffusion process is of importance for the understanding of the surface structure and its influence on a number of processes, as adsorption/desorption, catalytic properties, and epitaxial growth.

The diffusion process itself is an activated process, i.e., the molecule has to "jump" from one site to another site separated from it by a barrier. In principle there is no problem in jumping over several sites, thereby increasing the effective diffusion rate [32]. In any case, the diffusion constant $D(T)$ is temperature dependent and a simple Arrhenius expression is often used to model the temperature dependence, i.e.,

$$D(T) = D_0 \exp(-E_0/k_B T) \tag{3.1}$$

Thus measured values of $D(T)$ as a function of temperature are plotted against $1/T$ in order to extract D_0 and activation energy E_0 [33].

A simple expression for the diffusion constant can be obtained using the *transition state* approach (see Appendix A). This theory gives the following expression for the rate constant for diffusion in the x-direction

$$k_{TST}^{diff}(T_s) = \frac{k_B T_s}{h} n_d \frac{Q_y^\dagger Q_z^\dagger}{Q_R^{diff}} \exp(-E_a/kT_s) \tag{3.2}$$

where n_d is the dimension of the diffusion process and hence $n_d = 2$ for surface diffusion. The partition function Q_R^{diff} is calculated for the "reactants," i.e., the atom/molecule placed in the well on the surface, $Q_y^\dagger$ and $Q_z^\dagger$ are partition functions for the transition state involving vibrations in the y and z directions, and finally E_a is the effective barrier, i.e.,

$$E_a = V_0 + \frac{\hbar}{2}\left(\sum_k \omega_k^\dagger - \sum_k \omega_k\right) \tag{3.3}$$

the barrier height V_0 being obtained from the potential corrected for the vibrational zero-point energy difference [130]. The vibrational frequencies at the transition state and in the well are denoted by $\omega_k^\dagger$ and ω_k respectively. For an atom, we would have ω_x, ω_y, and ω_z for the three vibrations in the well, and $\omega_y^\dagger$ and $\omega_z^\dagger$ at the transition state it we consider a diffusion in the x direction. A diatomic molecule would have six frequencies when bound to the surface, and five at the transition state. Thus one of the vibrations (corresponding to the imaginary frequency) is replaced by a one-dimensional translation, which brings the atom/molecule over the barrier (see also Appendix A). The connection between the rate constant and the diffusion constant can be obtained using the random-walk model for the diffusion process [130]. Thus we have

$$D(T_s) = \frac{a^2}{2n_d} k_{TST}^{diff}(T_s) \tag{3.4}$$

where a is the jump length, i.e., the distance (in the diffusion direction) between the holes on the surface. The transition state model can be used not only for treating surface diffusion processes but also for other activated processes as adsorption or desorption processes (see, e.g., [29]). Usually just the "classical" version of transition state theory is used. However, recently an exact version of the theory has appeared—namely the *quantum transition state* theory. This approach has also been used for surface dynamical processes by Saalfrank and Miller [30], using, however, a simplified model potential energy surface (see also section 4.1). Returning to the classical version of transition state theory, it should be mentioned that it overestimates the rate due to the neglect of recrossing of the barrier, but it is possible to correct the theory so as to account for recrossings by calculating the transmission correction factor to the transition state result. This is done by running classical trajectories started at the transition state forward and backward in time. The trajectories are sampled from an equilibrium distribution at the transition state [156,31]. In this manner a dynamical correction factor can be obtained with a modest amount of numerical work.

If the rate constants for jumping from site i to j, k_{ij} is known, they can be inserted in a rate equation for the probability distribution $P_i(t)$ at a given site as a function of time. Thus we have

$$\frac{d}{dt}P_i(t) = -\sum_j k_{ij}P_i(t) + \sum_{l \neq i} k_{li}P_l(t) \tag{3.5}$$

Due to detailed balance ($k_{ji} = k_{ij}$), the transition matrix is a symmetric matrix. The above equations can easily be solved numerically, but a simple analytical solution can be obtained if the atoms/molecules only are allowed to jump to neighbor sites, i.e., from i to $i \pm 1$ with the same rate constant $k(T)$. Thus we have

$$\frac{d}{dt} P_i(t) = k(T)(P_{i+1} + P_{i-1} - 2P_i(t)) \tag{3.6}$$

which has the solution

$$P_n(t) = \exp(-2k(T)t) I_n(2k(T)t) \tag{3.7}$$

where I_n is a Bessel function. We notice that $\sum_n P_n(t) = 1$ and that the rate constant can be related to the diffusion constant by

$$D(T) = \frac{k(T)a^2}{2n_d} \tag{3.8}$$

where a, is the jump length.

In order to probe the surface and atom/molecules adsorbed at surfaces one can scatter test particles from the surface and measure their energy loss and angular distribution. The test particles which have been used are neutrons and He-atoms, which at low energies have de Broglie wavelength of the right magnitude to probe surface processes. This latter development has been primarily pioneered by Toennies and coworkers.

In order to extract information from angle-resolved experimental scattering data we need an expression for the so-called double-differential cross sections. If the energy transfer to the crystal and the adsorbates is small, the differential cross section can be approximated by using the Born approximation (see section 2.4). The model described below has been developed for neutron scattering but it has also been used for scattering of light atoms such as He from solid surfaces. In this approximation the scattering amplitude from N identical particles (see Fig. 3.1) from the initial state $\exp(i\mathbf{k}_0 \cdot \mathbf{R}) \mid i\rangle$ to final state $\exp(i\mathbf{k} \cdot \mathbf{R}) \mid f\rangle$ is approximated by (see also section 2.4)

$$\int d\mathbf{r} \langle i \mid \exp(i\mathbf{k}_0 \cdot \mathbf{R}) V(r) \exp(-i\mathbf{k} \cdot \mathbf{R}) \mid f\rangle \tag{3.9}$$

where $\mathbf{k}_0$ and $\mathbf{k}$ are the initial and final translational wavevectors of the scattered particles (the He atoms or neutrons). $\mid i\rangle$ and $\mid f\rangle$ are the initial and final state functions of the N particles and $V(r)$ the interaction potential. The total scattering cross section is obtained by summing over all final states f and averaging over initial states, which are assumed to have the probability distribution p_i.

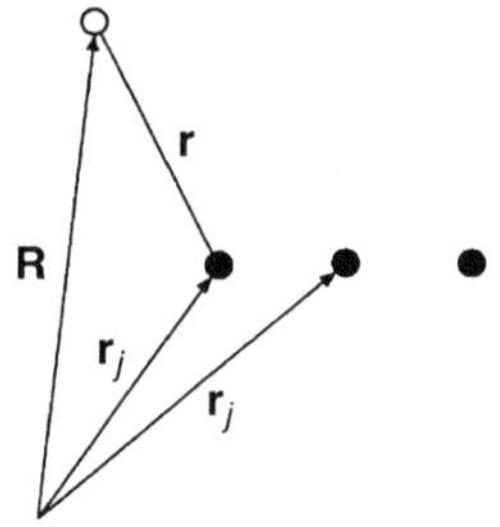

FIGURE 3.1 A particle with mass m located at **R** is scattered by N identical particles located at $\mathbf{r}_j$ ($j = 1, \ldots, N$).

Within the Born approximation the differential cross section per unit solid angle Ω and energy ε of the scattered particles can be written as [36]

$$\frac{d^2\sigma}{d\Omega d\varepsilon} = \frac{a^2 k}{k_0} S(\Delta\mathbf{K}, \omega) \tag{3.10}$$

where $a^2 = 2m^3/(2\pi)^2\hbar^6$ is the scattering length squared, and m is the mass of the scattered particle. The momentum transfer to the solid/adsorbates is (see Chapter 2)

$$\hbar\Delta\mathbf{K} = \hbar(\mathbf{k}_0 - \mathbf{k}) \tag{3.11}$$

and the function $S(\Delta K, \omega)$ is defined by

$$S(\Delta K, \omega) = W(\Delta K) \sum_i p_i \sum_f |\langle i| \sum_{j=1}^{N} \exp(i\Delta\mathbf{K}\cdot\mathbf{r}_j)|f\rangle|^2$$
$$\delta\left(k^2 - k_0^2 + \frac{2m}{\hbar^2}(E_f - E_i)\right) \tag{3.12}$$

where we have used the fact that (see Fig. 3.1) $\mathbf{R} = \mathbf{r} + \mathbf{r}_j$, p_i is the probability to find state i and

$$W(\Delta K) = \left| \int d\mathbf{r} V(r) \exp(i\Delta\mathbf{K}\cdot\mathbf{r}) \right|^2 \tag{3.13}$$

The last factor in eq. (3.12) ensures that the summations run over conditions obeying the energy conservation.

It is now convenient to use the fact that the delta-function can be written as

$$\delta\left(k^2 - k_0^2 + \frac{2m}{\hbar^2}(E_f - E_i)\right) = \frac{\hbar}{2m}\delta\left(\omega + \frac{E_i - E_f}{\hbar}\right)$$
$$= \frac{\hbar}{2m}\frac{1}{2\pi}\int_{-\infty}^{\infty} dt \exp(-i(\omega + (E_f - E_i)/\hbar)t) \tag{3.14}$$

where ω is defined below. The differential cross section can then be written as

$$\frac{d^2\sigma}{d\Omega d\varepsilon} = AS(\Delta K, \omega) \tag{3.15}$$

where

$$A = \frac{Nm^2}{4\pi^2\hbar^5}\frac{k}{k_0}W(\Delta K) \tag{3.16}$$

and

$$S(\Delta K, \omega) = \frac{1}{2\pi N}\int_{-\infty}^{\infty} dt \exp(-i\omega t)\sum_{ij}\langle \exp(-i\Delta\mathbf{K}\cdot\mathbf{r}_i(0))\exp(i\Delta\mathbf{K}\cdot\mathbf{r}_j(t))\rangle \tag{3.17}$$

where we have used the fact that

$$\sum_f \langle i|O(0)|f\rangle\langle f|O(0)|i\rangle \exp\left(\frac{i}{\hbar}(E_f - E_i)t\right)$$

$$= \sum_f \langle i|O(0)|f\rangle\langle f|\exp\left(\frac{i}{\hbar}H_0 t\right)O(0)\exp\left(-\frac{i}{\hbar}H_0 t\right)|i\rangle = \langle i|O(0)O(t)|i\rangle$$

and the closure relation $\sum_f |f\rangle\langle f| = 1$. The hamiltonian for the N particle system is H_0 with energies E_i such that $H_0 \mid i\rangle = E_i \mid i\rangle$. Furthermore the operator

$$O(0) = \sum_j \exp(i\Delta\mathbf{K}\cdot\mathbf{r}_j) \tag{3.18}$$

has been introduced. Likewise we see that the operator $O(t)$ is obtained as

$$O(t) = \exp(iH_0 t/\hbar)O(0)\exp(-iH_0 t/\hbar) \tag{3.19}$$

The reflection coefficient $S(\Delta K, \omega)$ can finally be written in a convenient form, namely as the Fourier transform in $\mathbf{r}$ and time t over a space-time correlation function $G(\mathbf{r}, t)$

$$S(\Delta\mathbf{K}, \omega) = \frac{1}{2\pi}\int\int \exp(i(\Delta\mathbf{K}\cdot\mathbf{r} - \omega t))\, G(\mathbf{r}, t)\, d\mathbf{r}dt \tag{3.20}$$

where ω is related to the energy transfer

$$\hbar\omega = \frac{\hbar^2}{2m}(k_0^2 - k^2) \tag{3.21}$$

By using that

$$\frac{1}{(2\pi)^3}\int d\mathbf{r}\exp(i(\Delta\mathbf{K} - \mathbf{K}\cdot\mathbf{r}) = \delta(\Delta\mathbf{K} - \mathbf{K}) \tag{3.22}$$

we see that the function $G(\mathbf{r}, t)$ is defined by (see also refs. [80,50,53])

$$G(\mathbf{r}, t) = \frac{1}{(2\pi)^3} \frac{1}{N} \sum_{ij} \int d\mathbf{K} \exp(-i\mathbf{K} \cdot \mathbf{r}) \langle \exp(i\mathbf{K} \cdot (\mathbf{r}_j(t) - \mathbf{r}_i(0))) \rangle \qquad (3.23)$$

which by using a definition of the delta function, also can be written as

$$G(\mathbf{r}, t) = \frac{1}{N} \left\langle \sum_{ij} \delta(\mathbf{r} + \mathbf{r}_i(0) - \mathbf{r}_j(t)) \right\rangle \qquad (3.24)$$

In other words, the function can be interpreted as the probability of finding the particle at $\mathbf{r}$ at time t given that it is at $\mathbf{r} = \mathbf{0}$ at $t = 0$.

Other models for surface diffusion assumes a random walk process which in the long-time limit [130] leads to a Gaussian distribution for $G(\mathbf{r}, t)$; i.e., for the two-dimensional surface diffusion we have (after many jumps)

$$G(\mathbf{r}, t) = \frac{1}{4\pi Dt} \exp(-r^2/4Dt) \qquad (3.25)$$

where D is the diffusion constant. Inserting this expression in eq. (3.20) we get

$$S(\Delta\mathbf{K}, \omega) = \frac{1}{i\omega + D(\Delta K_x^2 + \Delta K_y^2)} \qquad (3.26)$$

Thus, if the energy loss ($\hbar\omega$) and momentum transfer in a given crystal direction ($\Delta\mathbf{K}$) can be obtained by time-of-flight detection of the probe particles, we can determine the diffusion constant. Considering the intensity, we get

$$I = |S(\Delta\mathbf{K}, \omega)|^2 = \frac{\hbar^2}{\hbar^2\omega^2 + D^2\hbar^2(\Delta K)^4} \qquad (3.27)$$

Thus the reflection coefficient for scattering from a diffusing particle has a Lorentzian energy shape with a width of

$$\Gamma = 2\hbar D(\Delta K)^2 \qquad (3.28)$$

Plots of the intensity as a function of the energy transfer therefore give information about the halfwidth [54].

Table 3.1 shows data obtained using He-atom scattering from adsorbed molecules. The experimental data are interpreted using the jump model of Chudley and Elliot [35]. In this model one expresses the broadening as

$$\Gamma = 4\hbar \sum_j \nu_j \sin^2\left(\frac{1}{2}\Delta\mathbf{K} \cdot \mathbf{j}\right) P_j \qquad (3.29)$$

TABLE 3.1 Diffusion Coefficients and Barriers Obtained by He-Scattering from Adsorbed Molecules on Various Surface

System	Direction*	E_0 meV	D_0 10^{-4} cm^2/sec
CO/Ni(110)	$\langle 1\bar{1}0\rangle$	57 ± 4	2.1 ± 0.70
	$\langle 001\rangle$	35 ± 4	0.7 ± 0.4
CO/Ni(001)	$\langle 100\rangle$	27 ± 3	0.5 ± 0.2
	$\langle 110\rangle$	33 ± 3	0.7 ± 0.2
CO/Cu(001)	$\langle 100\rangle$	32 ± 6	~ 0.1
Na/Cu(001)	$\langle 100\rangle$	53.5 ± 1.4	0.7 ± 0.3
	$\langle 110\rangle$	54 ± 6	1.8 ± 0.8

* As mentioned in Chapter 2 this notation is used to specify a specific crystallographic direction. Data from ref. [33].

where ν_j is the frequency for a jump in direction $\mathbf{j}$ with length $|\mathbf{j}|$ and P_j the probability for a certain jump type. As in the transition state theory, the jump frequency can often be approximated by a simple activated process

$$\nu_j = \nu_j^0 \exp(-E_b/k_B T_s) \tag{3.30}$$

where E_b is the diffusion barrier and T_s the surface temperature. For single jumps $(j = 1)$ between neighbor cells a distance a apart we obtain

$$\Gamma = 4\hbar\nu^0 \exp(-E_b/k_B T_s) \sin^2\left(\tfrac{1}{2}\Delta K a\right) \tag{3.31}$$

We notice that for small values of ΔK we have

$$\Gamma \sim \hbar\nu^0 a^2 (\Delta K)^2 \exp(-E_b/k_B T_s) \tag{3.32}$$

which by using eq. (3.28) gives

$$D = a^2 \nu^0 \exp(-E_b/k_B T_s) \tag{3.33}$$

which is the same expression obtained from eqs. (3.2) and (3.4) with the high temperature approximation for the vibrational partition functions $Q \sim k_B T/h\nu$. From TST we obtain $\nu^0 = \nu_x \nu_y \nu_z / \nu_y^{\dagger} \nu_z^{\dagger}$ as an effective attempt frequency for jumping over the barrier. There is experimental evidence that a large fraction of the jumps of which the activated diffusion process consists may actually be double jumps rather than single jumps [33].

3.1 MOLECULAR DYNAMICS CALCULATIONS

The diffusion constant can be calculated through molecular dynamics simulations using the Einstein relation,

$$D = \lim_{t \to \infty} \frac{1}{2tn_d} \langle \Delta \mathbf{R}^2 \rangle \tag{3.34}$$

where n_d is the dimensionality of the diffusion process, i.e., $n_d = 2$ for surface diffusion, the brackets indicate an ensemble average, and $\Delta\mathbf{R}$ the displacement vector of the center of mass of the molecule. The diffusion constant can also be obtained through the velocity autocorrelation function [55]

$$D = \frac{1}{n_d} \int_0^{\infty} dt \langle \mathbf{v}(0) \cdot \mathbf{v}(t) \rangle \tag{3.35}$$

i.e., as the time-averaged correlation function of the initial velocity vector and the velocity vector at time t. Again the brackets indicate an ensemble average. Thus introducing the two-dimensional momentum distribution and $n_d = 2$, we have

$$D(T) = \frac{1}{8\pi mkT} \int dp_x \int dp_y \frac{\Delta X^2 + \Delta Y^2}{t} \exp(-(p_x^2 + p_y^2)/(2mkT)) \tag{3.36}$$

where m is the mass of the molecule. In ref. [56] the diffusion coefficients for hydrogen on a Cu(111) surface was calculated using both the Einstein and the Green-Kubo relation (3.35). The results are given in Table 3.2.

The calculated numbers show an increase with temperature, which at low temperatures may be fitted to an Arrhenius type expression. The barrier depends, however, as we have seen on the "direction" of the diffusion process and hence if the initial diffusion direction is chosen at random only some average will be obtained. If the temperature is high such that a majority of molecules have energies above the "barrier," then the Arrhenius-type behavior cannot be expected. This is also clearly seen from the numbers in the Table 3.2. The diffusion constant approaches a limiting value at high temperatures.

TABLE 3.2 Diffusion Constant $D(T)$ in cm^2/sec Obtained in [56] for Physisorbed Hydrogen Diffusion on a Cu(111) Surface

Temperature/K	Eq.(3.34)	Eq. (3.35)
50	0.00090	0.00087
80	0.00273	0.00264
100	0.00347	0.00336
120	0.00381	0.00369
150	0.00386	0.00374

EXERCISES

3.1: Use the fact that $\int_{-\infty}^{\infty} dx \cos(px) \exp(-qx^2) = \sqrt{\pi/q} \exp(-p^2/(4q))$ to verify eq. (3.26).

3.2: Obtain eq. (3.35) from eq. (3.34).

3.3: After correction with a Debye-Waller factor (see section 8.4), the following Lorentzian halfwidths could be determined from time-of-flight spectra of He-atoms scattered from a surface covered with diffusing CO-molecules.

Temperature (K)	Γ (meV)
220	0.22
240	0.28
260	0.35
280	0.41

Determine the diffusion coefficient in cm^2/sec and the activation energy for the diffusion process E_0 if $\Delta K = 1.1$ Å^{-1}.

3.2 QUANTUM DIFFUSION

Quantum diffusion is important for low mass molecules or atoms which may tunnel through the barrier for diffusion. The tunneling process can be described either exactly, using a wavepacket representation of the atom, or approximately, using semiclassical tunneling corrections known from gas-phase reactions. Thus the rate constant for the process can be obtained through the transition state theory corrected with a tunneling factor [156]. The simplest tunneling correction factor is the one due to Wigner [34]

$$\kappa = 1 + \frac{1}{24} \frac{(\hbar|\omega^{\kappa}|)^2}{(kT_s)^2} \tag{3.37}$$

where ω^{κ} is the imaginary freqency at the transition state for the motion in the direction of diffusion. The effect of tunneling can also be included by using an improved version of transition state theory, *canonical variational* transition state theory. This theory has been used to study surface diffusion by Truhlar and coworkers [44].

Quantum tunneling has been considered using the flux-flux correlation function and a Monte-Carlo path integral method by Jaquet and Miller [45]. Metiu et al. have considered quantum diffusion of hydrogen atoms also using the flux-flux correlation method [46].

In the flux-flux correlation method [47] the rate constant is obtained as

$$k = \lim_{t\to\infty} \frac{g}{Q_r} \int_0^t dt' C_{ff}(t') \tag{3.38}$$

where g is a symmetry factor which indicates the number of equivalent diffusion paths. Q_r the partition function for the "reactants," and $C_{ff}(t)$ the flux-flux correlation function for diffusion over a dividing surface, which for diffusion along the x-axis is defined by a plane perpendicular to the axis at $x = x_s$. In this case the flux operator is obtained as

$$F_x = \tfrac{1}{2m}[\hat{p}_x \delta(x - x_s) + \delta(x - x_s)\hat{p}_x] \tag{3.39}$$

where m is the mass of the diffusing particle and $\hat{p}_x$ a momentum operator. The correlation function is obtained as

$$C_{ff}(t) = \frac{d}{dt}\mathrm{Tr}(F_x \exp(iHt_c^*/\hbar)F_x \exp(-iHt_c/\hbar)) \tag{3.40}$$

where Tr denotes the quantum mechanical trace, $t_c = t - i\beta\hbar/2$, and $\beta = 1/k_B T$. The partition function is given as the trace over the Boltzmann operator

$$Q_r = \mathrm{Tr}(A \exp(-\beta H)) \tag{3.41}$$

where [46] the trace over the operator A gives the size of the unit cell in which the atom/molecule is located initially. Thus for a unit cell with lattice constant a we have

$$A = \int_{-a}^{0} dx \int_{-a/2}^{a/2} dy |xy\rangle\langle xy| \tag{3.42}$$

where $| xy\rangle$ denote a normalized two-dimensional wavefunction in the surface coordinates x, y (for instance a product of free particle wavefunctions). With this expression for the operator we have $\mathrm{Tr}(A) = a^2$.

The coupling to the lattice motion is of importance for the diffusion process. This was accounted for by Metiu et al [46] by treating the lattice atom dynamics classically, i.e., a quantum-classical model was introduced. The simplest way of coupling the two systems, the molecule and the surface, together is obtained by introducing a "mean field" approach in which the quantum system feels an average lattice position — averaged over an ensemble of trajectories according to a given lattice temperature. The lattice motion influences the diffusion process through several factors, the presence of zero-point energy is of importance, and the phonon motion may enhance the diffusion process by shortening the diffusion path through what has been called phonon-assisted tunneling motion [44, 260].

At the simplest level of the discussion, the coupling between the diffusion process and phonon excitation/deexcitation processes can be thought of as adding a time-dependent frictional or dissipative force to the equations of motion for the diffusing atom/molecule. Through the generalized Langevin equation is it possible to obtain the frictional force from the fluctuation-dissipation theorem, which relates the autocorrelation function of a fluctuating force to the dissipative (see Chapter 7). Aside from the role played by the phonons, excitation of electronhole pairs is also a possible energy loss mechanism which will add additional friction to the diffusion process. This may also be modeled by the insertion of an electronic friction term in the Langevin equation [258]. Electronic friction will be considered in Chapter 11.

EXERCISE

3.4: Determine the diffusion constant D_0 and the activation energy E_b from the following data (adapted from [48]) for hydrogen diffusing on Pt(111). The table gives values of Γ in units of 10^{-6} eV as a function of temperature and ΔK in Å^{-1}.

ΔK	T = 140 K	T = 200 K	T = 250 K
−1.0	34	127	211
−1.5	34	135	230
−2.0	15	88	122

Use a single-jump model with jump length $j = a\cos(30°)$ and $a = 2.77$ Å.

4

INTERACTION POTENTIALS

4.1 ATOM-SURFACE INTERACTION

As mentioned in Chapter 1, simple models for atom/molecule-surface interaction were already suggested by Lennard-Jones and London [111]. Starting with the Lennard-Jones 12-6 potential for atom-atom interaction

$$V(r) = C((r_0/r)^{12} - 2(r_0/r)^6) \tag{4.1}$$

where r is the atom-atom distance and by summing over the solid atom positions (see exercise 4.1) we obtain the 9-3 potential mentioned in the introduction. The coefficient in the Lennard-Jones potential $C' = 2Cr_0^6$ is known theoretically through the London formula

$$C' = \frac{3}{2}\frac{I_A I_B}{I_A + I_B}\alpha_A \alpha_B \tag{4.2}$$

where I_A and I_B are the ionization potentials of the two atoms and α_A and α_B the polarizabilities. From known values of these quantities we can predict the surface binding energy and vice versa (see exercises). In gas-phase work the dispersion r^{-6} term is usually damped at short range with a damping function. A popular choice is to use the one due to Tang and Toennies [112]. This term has also been used when constructing atom-surface potentials [40,41,42] and hence the long range atom-atom potential takes the form

$$-C'r^{-6}\left(1 - \exp(-br)\sum_{k=0}^{6}\frac{(br)^k}{k!}\right) \tag{4.3}$$

where the b parameter can be used to monitor the position of the physisorption well. But often it is just taken as the steepness parameter in the short-range repulsive potential $A\exp(-br)$.

Extensive information on weakly bound, physisorbed atoms on solids is available from bound state resonance data obtained from beam scattering experiments. Strong changes in the scattering intensities are observed when selective adsorption in the physisorption well occurs (see, for instance, ref. [113]). The resonance positions yield the eigenvalues E_n of the potential $V(z)$. These satiesfies to a good approximation the Bohr-Sommerfeld quantization condition

$$\hbar\pi\left(n+\frac{1}{2}\right)=\oint dz p(z) \tag{4.4}$$

where $p(z)=\sqrt{2m(E_n-V(z))}$. The analytical form to which many beam data may be fitted is

$$V(z)=D\exp(-a(z-z_0))-C_3/z^3 \tag{4.5}$$

where the value z_0 is chosen such that $V(z_0)=0$. Introducing also the value z_m where the potential has its minimum, i.e., $(dV(z)/dz)|_{z=z_m}=0$ and the dimensionless variables $x=z/z_0$, $z_0=(C_3/D)^{1/3}$ the exp-3 potential can be written as

$$f_{exp-3}=D\left(\frac{3}{u-3}\exp\left(-u\left(\frac{x}{x_m}-1\right)\right)-1/x^3\right) \tag{4.6}$$

where $x_m=(1-3/u)^{1/3}$, $x_m=z_m/z_0$, and $u=az_m$ [114]. D is the well depth and C_3 the long-range parameter. Thus in these reduced variables the potential takes a "universal" form $v_0(x)=Df(x)$.

Table 4.1 gives typical values for inert gas-graphite and Ag(111) interaction potentials. Accurate estimates of many rare gas surface potentials are available in the literature. For an accurate Xe-Pt(111) potential see, for instance, [121].

TABLE 4.1 Atom-Surface Parameters for Rare Gas Atoms Interacting with Graphite and Ag(111) Surfaces

System	C_3 meV Å^3	D meV	z_0/Å	z_m/Å
He-Graphite	185	16.4	2.24	2.73
Ne-	346	32.9	2.19	2.82
Ar-	1208	77.1	2.50	3.14
Kr-	1730	98.9	2.60	3.24
Xe-	2460	132.9	2.65	3.49
He-Ag(111)	249	4.95	3.69	4.48
Ar-	1620	74	2.80	3.69
Kr-	2260	112	2.72	3.30
Xe-	3380	172	2.70	3.55

Data from ref. [110].

We note that the long-range part varies as z^{-3}, which may be rationalized by a dipole image-dipole interaction, or from an atom-atom dispersion interaction r^{-6} summed over the lattice atoms (see exercise 4.1). Often the potential is modeled by an expression of the type

$$V(z) = A\exp(-az) - \frac{C}{(z - z_m)^3} \tag{4.7}$$

where z_m is the distance to the "Jellium" edge, located above the first layer of atoms. A jellium denotes a homogeneous electron gas (or liquid). Typical values for z_m are about 1Å (see, for instance, ref. [119]).

The physisorption wells are in the meV range and the atoms are physisorbed at some value of the distance $z \sim 2$–4 Å from the surface. For a discussion of the systematic trends in atom-surface potentials, see also ref. [120].

The simple one-dimensional potential $V(z)$ can be used to illustrate what happens if a molecule is scattered from a metal surface. At a distance of about 2–3 Å above the surface, the molecule is bound in a physisorption well. The energy which is needed to dissociate the molecule is large in the gas phase. However, if charge can be transferred from the metal to the antibonding levels of the molecule, the bond is weakened and the dissociation energy drops. The workfunction ϕ is the energy needed to remove an electron from the Fermi-level of the metal (see Table 4.2). When the molecule recieves the electron, we gain the electron affinity A. As the molecule approaches the surface, the interaction between the electrons in the metal and those of the molecule changes the electron affinity by effectively lowering the molecular antibonding energy level. At some distance, the workfunction and the electron affinity are equal (see Fig. 4.1). At this distance the dissociation of the molecule requires that the available energy is above the barrier. This picture is, however, only qualitatively correct. A more realistic picture shows a contour plot of the potential energy surface as a function of the bond-distance r and the surface distance z (Fig. 4.2).

TABLE 4.2 The workfunction ϕ in eV for Some Metals

Metal	ϕ	Metal	ϕ	Metal	ϕ
Au(111)	5.31	Au(110)	5.37	Au(100)	5.47
Cu(111)	4.94	Cu(110)	4.48	Cu(100)	4.59
Ag(111)	4.74	Ag(110)	4.52	Ag(100)	4.64
Al(111)	4.24	Al(110)	4.06	Al(100)	4.41
Ni(111)	5.35	Ni(110)	5.04	Ni(100)	5.22
W(111)	4.47	W(110)	5.25	W(100)	4.63
Re(1011)	5.75	Pt(111)	5.7	Fe(100)	4.67
Fe(111)	4.81	Na	2.75	K	2.30

Data from ref. [122]. The workfunction depends on coverage and also on the surface structure.

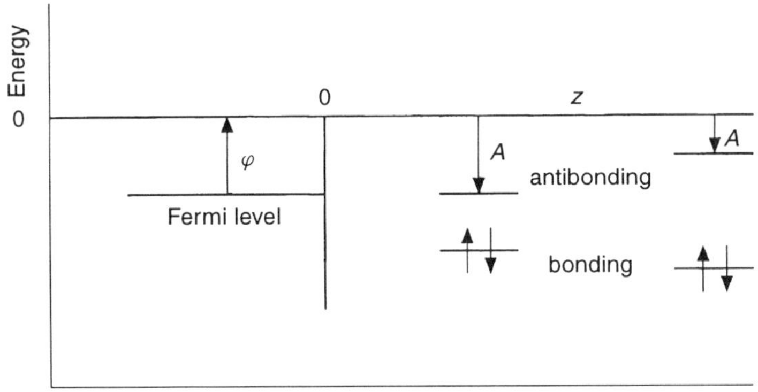

FIGURE 4.1 The work function ϕ is the energy needed to remove an electron from the metal. The energy released when the electron enters the antibonding level of the molecule is the affinity *A*. The affinity increases as the molecule approaches the surface.

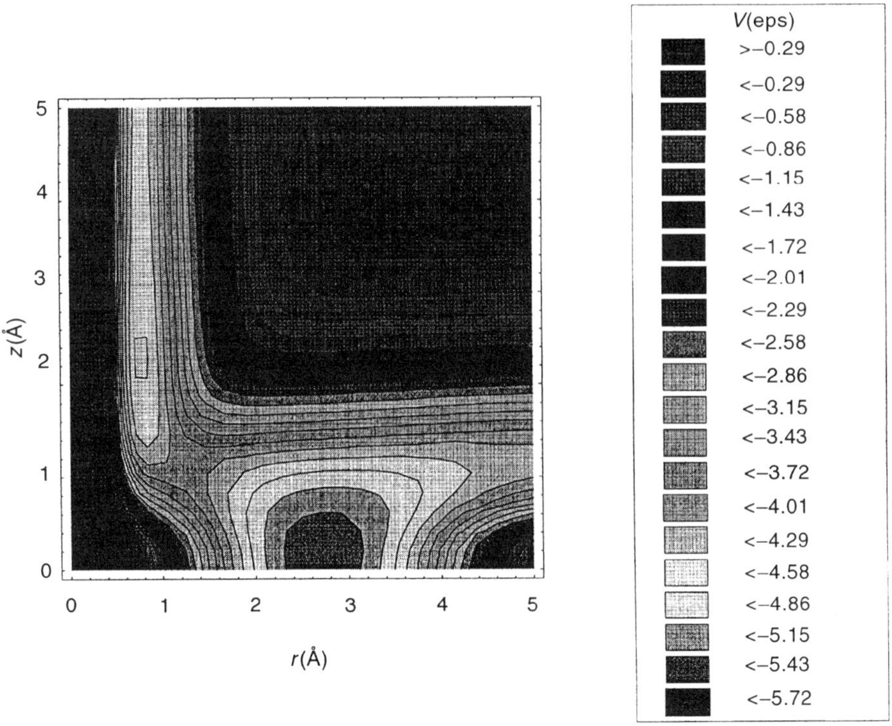

FIGURE 4.2 Two-dimensional surface contour map of the interaction between a molecule and a metal surface. The bond distance is *r* and the distance from the center of mass to the surface is *Z*. The unit of energy on contour maps is eps = 100 kJ/mol.

We notice that the molecule can pass either an early barrier located in the entrance channel (where $r \sim r_e$ the gas phase equilibrium distance) or a late barrier in the exit channel where $r > r_e$. If the barrier is late, then vibrational excitation of the incoming molecule enhances the dissociative sticking process; if it is early, the dissociation is enhanced by increasing the translational energy of the molecule. This early and late barrier discussion is therefore identical to the one known from gas-phase reaction dynamics [123]. Thus it is natural to try to use similar model potentials in molecule surface interaction to those which have been used in gas-phase dynamics. Such a model potential is the one due to London, Eyring, Polanyi, and Sato, in short, denoted the LEPS potential.

EXERCISES

4.1. Show that an atom-atom long-range interaction

$$V_{a\alpha} = -C_6/r^6 \tag{4.8}$$

between the gas-atom a and the surface atom α when summed over α has the behavior

$$v_0(z) = -\frac{\pi}{6}\frac{C_6}{z^3}\frac{1}{abc} \tag{4.9}$$

where a, b, and c are the lattice constants in the x, y, and z directions. *Hint*: Express the position of a lattice atom by $(x' + ma, y' + nb, -lc)$, where m and n take values from $-\infty$ to $+\infty$ and l from 0 to ∞. Use that

$$\int_0^\infty \frac{x^{\mu-1}dx}{(1+\beta x^p)^\nu} = \frac{1}{p}\beta^{-\mu/p}B(\mu/p, \nu - \mu/p) \tag{4.10}$$

and $B(x, y) = \Gamma(x)\Gamma(y)/\Gamma(x + y)$.

4.2. Show that the C_9 constant in the potential

$$V(z) = C_9 z^{-9} - C_3 z^{-3} \tag{4.11}$$

is given as

$$C_9 = \frac{\pi}{45}\frac{1}{abc}C_{12} \tag{4.12}$$

and that the minimum of this potential is given by

$$z_0 = (1/5)^{1/6} r_0 \tag{4.13}$$

where r_0 is the minimum of the atom-atom 12-6 potential (eq. (4.1)). Finally show that the surface binding energy of the 9-3 potential is related to the atom-atom binding energy D by the formula

$$V_b = -V(z_0) = \frac{\pi}{abc} D r_0^3 \frac{2}{9}\sqrt{5} \tag{4.14}$$

4.2 SEMI-EMPIRICAL MODELS

The strong almost covalent, bonds formed between many atoms or molecules and metal surfaces can be explained phenomenologically by electron transfer from the metal to the molecule [153]. The electron transfer can, as mentioned, happen at a distance from the surface where the electron affinity A of the atom/molecule exceeds the workfunction ϕ of the metal. This is shown schematically in Figure. 4.1. In this model we asymptotically have a potential energy surface for neutral-neutral interaction, a curve crossing at $R = R_c$, and a negative ion positive surface interaction at $R < R_c$. At short distances, overlap between electron clouds will give rise to a repulsive interaction. When fitting the potential surface parameters to experimental data on surface-bound molecules or to electronically relaxed ab initio data, one models the lower adiabatic interaction. A process as charge transfer through electron redistribution has, therefore so to speak, taken place; i.e., a dynamical process is in fact incorporated in the "static" potential surface for the nuclear motion. If the atoms move infinitely slowly, then this "Born-Oppenheimer approach" is believed to be adequate.

Since accurate ab initio calculations on atom/molecule-surface potentials are scarce, one will often have to use semi-empirical or empirical models in order to obtain the many-body interaction potential necessary for modeling chemical reactions at surfaces. However, a similar situation is encountered in gas-phase chemistry, and although more accurate calculations are now appearing, it has been quite popular also to use semi-empirical methods in this area. The most popular of these methods is the so-called LEPS method, which originally was formulated for the H_3 3-electron system by London, Eyring, and Polanyi with a modification due to Sato [131]. The expression involves Coulomb and exchange integrals for the diatomic fragment pairs. In Appendix B the valence bond theory, which gives the results for four-and three-electron systems, is reviewed.

For a three-electron valence bond structure one obtains the energy:

$$E = Q - \left[\frac{1}{2}((J_{AB} - J_{AC})^2 + (J_{AB} - J_{BC})^2 + (J_{BC} - J_{AC})^2)\right]^{1/2} \tag{4.15}$$

where

$$Q = Q_{AB} + Q_{AC} + Q_{BC} \tag{4.16}$$

Q is the coulombic binding energy between the three atoms A, B, and C and J_{AB}, J_{AC} and J_{BC} are the corresponding exchange integrals. Removing the electron c to infinity we get

$$E_{AB} = Q_{AB} \pm J_{AB} \tag{4.17}$$

where the upper sign is the bonding state and the lower sign the antibonding state (since $J_{AB} < 0$). A modification due to Sato introduces the overlap integrals such that the above equation is modified to:

$$E_{AB}^{\pm} = \frac{Q_{AB} \pm J_{AB}}{1 \pm S_{AB}^2} \tag{4.18}$$

The above result is now used semi-empirically by identifying E_{AB}^{+} with the experimentally obtained bonding and E_{AB}^{-} with the antibonding states. Usually these are taken to be Morse and anti-Morse potentials, i.e.,

$$E^{+} \to D_e(\exp(-\beta(r - r_e) - 2)\exp(-\beta(r - r_e)) \tag{4.19}$$

$$E^{-} \to \tfrac{1}{2}D_e(\exp(-\beta(r - r_e) + 2)\exp(-\beta(r - r_e)) \tag{4.20}$$

where D_e is the potential well depth, r_e, the equilibrium distance, and β are obtainable from spectroscopic measurements (see, for instance, Herzberg [132]).

For a four-electron system (Fig. 4.3) we would get:

$$\begin{aligned} E = {} & Q_{AB} + Q_{BC} + Q_{AC} + Q_{BD} + Q_{AD} + Q_{CD} \\ & - \left[\tfrac{1}{2}((\alpha - \beta)^2 + (\beta - \gamma)^2 + (\gamma - \alpha)^2)\right]^{1/2} \end{aligned} \tag{4.21}$$

where Q_{AB} is the Coulomb energy for the two-electron system as before and

$$\alpha = J_{AB} + J_{CD}, \tag{4.22}$$

$$\beta = J_{AD} + J_{BC} \tag{4.23}$$

and

$$\gamma = J_{AC} + J_{BD} \tag{4.24}$$

In order to modify this expression to the diatom-surface case it has been suggested that one should treat the solid as an extended two-electron source (see Fig. 4.3),

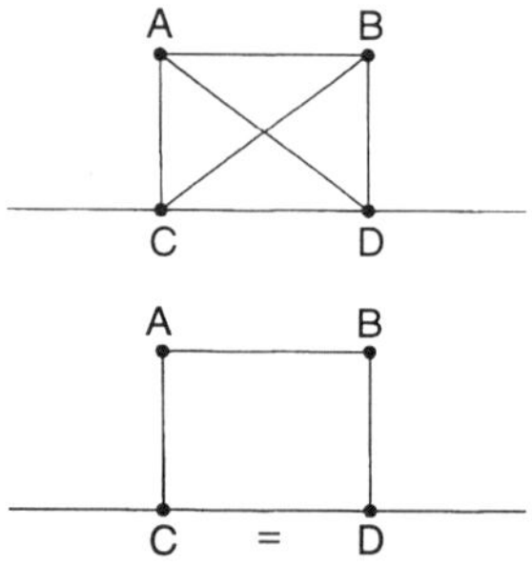

FIGURE 4.3 The valence bond structures for a molecule a-b interacting with a metal are obtained by regarding the metal as an extended electron source of the pseudo-molecule c-d.

omitting from the valence bond structure those terms in (4.21) denoted by A-C, B-D, and C-D. Thus we get

$$E = Q_{AB} + Q_{CB} + Q_{AD} - \left[\tfrac{1}{2}(J_{AB} - \beta)^2 + \beta^2 + J_{AB}^2)\right]^{1/2} \tag{4.25}$$

From eqs. (4.18),(4.19),(4.20) we obtain

$$Q = \tfrac{1}{4}D[(3 + \Delta)\exp(-2\beta(r - r_e)) - (2 + 6\Delta)\exp(-\beta(r - r_e)) \tag{4.26}$$

and

$$\alpha = \tfrac{1}{4}D[(3\Delta + 1)\exp(-2\beta(r - r_e)) - (6 + 2\Delta)\exp(-\beta(r - r_e))] \tag{4.27}$$

The parameters entering these equations are then the Sato parameter $\Delta = S_{AB}^2$, A-B Morse parameters, and the A-surface and B-surface interaction. The above scheme could be extended to three atom-surface interaction by having five centers instead of four [133]. For a recent attempt to use an LEPS-like functional form for obtaining an analytical expression for the hydrogen-Cu(111) surface, see ref. [134].

The atom-surface interaction is often approximated by [135]

$$V_{A-S} = D(X, Y)[\exp(-2\beta_{A-S}(z - z_0) - 2\exp(-\beta_{A-S}(z - z_0))] \tag{4.28}$$

where we have indicated that the dissociation energy D for the surface-bound atom depends upon the binding site (X, Y). In general we have for the dissociation, the slope, and the distance parameters

$$D = D(X, Y) \tag{4.29}$$

$$\beta = \beta(X, Y) \tag{4.30}$$

and

$$z_0 = z_0(X, Y) \tag{4.31}$$

If the corrugation of the surface is weak we may use the expansion

$$D = D_0(1 + \delta[\cos(2\pi x/a) + \cos(2\pi y/a)] + \cdots) \tag{4.32}$$

and keep only a few terms in the expansion. The parameters D_0, δ, β_{A-S}, and z_0 may be estimated from experimental information on binding energies, positions, and frequencies. The method described above does include the surface structure in the potential. But it neglects variation in the potential due to phonon excitation, i.e., motion of the lattice atoms.

One simple way of incorporating this dependence in the LEPS-surface has been introduced by DePristo et al. for the N_2-W [150] and H_2-Ni [151] interactions by fitting the three Morse parameters D_{AS}, R_{AS}, and α_{AS} to ab initio data [152]. The distance parameter R_{AS} is expressed as

$$R_{AS} = \left(\frac{3}{4\pi n(\mathbf{X}_A)}\right)^{1/3} \tag{4.33}$$

where $n(X)$ is the electron density at the distance X from the surface to the atom. This density is obtained by summing over the contribution from each surface atom, i.e.,

$$n(\mathbf{X}_A) = \sum_{\alpha} n(R_{A\alpha}) \tag{4.34}$$

where $R_{A\alpha}$ is the distance from surface atom α to the atom A. The atomic densities are obtained as a sum of contributions from s and d valence electron contributions [150]. This model therefore similar in spirit to the EDIM (embedded diatomics in molecules) model described below. Another possible simple way of including surface motion, i.e., the method which uses the fewest parameters, is to write the interaction as a sum of atom-atom pair potentials. Thus we assume that

$$V_{A-S} = \sum_{i=1}^{N_1} V_{Ai}(R_{Ai}) \tag{4.35}$$

where R_{Ai} is the distance from the atom A in the gas phase to the ith atom in the solid. If the atom A is unable to migrate below the top layer of the solid the summation (4.35) could run over the top-layer atoms only. A priori, one would not expect an atom-atom potential to be adequate to describe a system where the electrons are so "diffuse" as in a metal. However, if (4.35) is not sufficiently accurate, one may add a three-body interaction term (see, e.g., [42,43])

$$V_3 = \sum_{i=1}^{N_1} \sum_{j=1}^{N_1} v_3(R_{Ai}, R_{Aj}) \tag{4.36}$$

and so on. The two-body interaction term V_{Ai} can again be approximated by a Morse-potential

$$V_{Ai} = \exp(-\beta_A R_{Ai})(D_A \exp(-\beta_A R_{AI}) - E_A) \tag{4.37}$$

In order to get the correct asymptotic behavior, one should, however, add a long-range attractive z^{-3} potential to eq. (4.35). Expression (4.37) contains 3 parameters: D_A, E_A and β_A. These parameters can be obtained from experimental information on binding site, energy, distance, and atom-surface vibrational frequencies. The same information could be used to determine the parameters D_B, E_B, and β_B for the B-surface interaction. For the molecule AB-surface interaction we need information at two binding sites in order to determine D_A^*, D_B^* and so on, where the asterisk indicate that the atom is placed in the molecule AB. Usually the two values D_A and D_A^* are different (see ref. [136]) and it is necessary to introduce a switching function which depends upon the A-B distance. Thus we may define

$$D_A(r_{AB}) = D_A^* f(r_{AB}) + D_A(1 - f(r_{AB})) \tag{4.38}$$

where the switching function should have the limits

$$f(r_{AB}) \to \begin{cases} 0 & \text{for } r_{AB} \to \infty \\ 1 & \text{for } r_{AB} \to r_{AB}^0 \end{cases} \tag{4.39}$$

where r_{AB}^0 is the A-B distance of the surface-bound AB molecule. The data in Table 4.3 for the vibrational frequencies of CO adsorbed on a Pt surface at two sites can, together with the binding energy at the sites, be used for a nonlinear least squares fitting to obtain an estimate of the potential parameters. If the molecule is relatively strongly bound to the surface, as is the case for CO on platinum, it is possible to study the vibrational modes by various experimental technique (LEED or HAS). We see (Fig. 4.4) that the molecule has six vibrational modes corresponding to a stretch of the C–O bond, the CO-surface bond, two hindered rotations, and translations. The latter being doubly degenerate at the top-site.

The total interaction potential is, for example, assumed to be of the form

$$V = V_{CO}(r_{CO}) + V_{CO-Pt} \tag{4.40}$$

where V_{CO} is the Morse gas-phase interaction

$$V_{CO} = D_e^{CO}(1 - \exp(-\beta_{CO}(r_{CO} - r_{CO}^e)))^2 \tag{4.41}$$

and V_{CO-Pt} is given by

$$V_{CO-Pt} = \sum_{i=1}^{N_1} \sum_{n=C,O} \exp(-\beta_n R_{ni})[D_n \exp(-\beta R_{ni}) - E_n) + B_n R_{ni} + C_n R_{ni}^2] \tag{4.42}$$

TABLE 4.3 Experimental Data for CO-Pt(111) Interaction

Mode	Frequency in cm^{-1} Top site	Mode	Bridge Site	Type
ω_1	2100	ω_1	1850	C–O vibration
ω_2	480	ω_2	380	Pt-CO vibration
ω_3	411	ω_4	298	Hindered rotation
ω_3	411	ω_5	360	Hindered rotation
ω_4	48.5	ω_3	60	Hindered translation
ω_4	48.5	ω_6	60	Hindered translation

Adapted from [137].

Some of the low-frequency values have recently [138] been revised slightly.

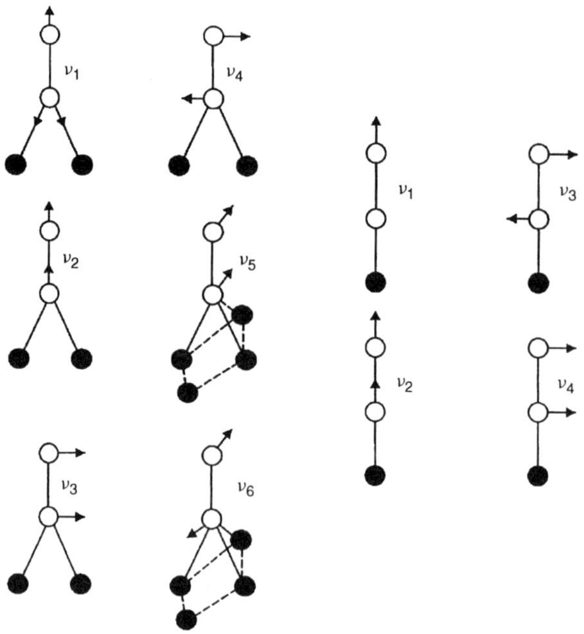

FIGURE 4.4 The six vibrational normal modes for a surface-bound diatomic molecule at the bridge and top sites.

i.e., the interaction potential contains 10 parameters (β_n, D_n, E_n, B_n, and C_n). The terms B_nR_{ni} and $C_nR_{ni}^2$ have been added to obtain additional flexibility in the fit. Also the two binding distances, i.e., the distance from the carbon and oxygen atoms to the surface, can be incorporated as parameters, i.e., a total of 12 parameters are then involved. The above expression does reduce to the gas-phase CO interaction potential for $R_{ni} \to \infty$. Experimentally it is known that the CO molecule is bound in a perpendicular configuration with the C-end downward, and the two binding energies are 131 kJ/mole at the top and 110 kJ/mole at the bridge site [139], with a C-distance to the surface of 1.85Å and 1.55Å respectively [140]. These data obtained from LEED experiments, are typically given with an experimental uncertainty of about 5%. The CO bond for the surface-bound potential is 1.15Å [140] compared with 1.128Å in the gas phase. The potential parameters can now be estimated by a nonlinear least squares minimization of the function

$$F(\mathbf{x}) = (V_{top}(\mathbf{x}) - V_t^*)^2 + (V_{bridge}(\mathbf{x}) - V_b^*)^2 + \sum_i w_i(\omega_i^{top}(\mathbf{x}) - \omega_i^t)^2 + \sum_i w_i(\omega_i^{bridge}(\mathbf{x}) - \omega_i^b)^2 + w_g \sum_{k=1,2} \left[\left(\frac{dV}{dx_k}\right)^2 + \left(\frac{dV}{dy_k}\right)^2 + \left(\frac{dV}{dz_k}\right)^2 \right] \tag{4.43}$$

where w_i and w_g are weights. The first two terms secure minimization of the binding energy to the values mentioned above, $V_t^* = -131$ kJ/mol and $V_b^* = -110$ kJ/mole. The next terms try to fit the frequencies given in Table 4.3 and the last term tries to enforce the potential to have a minimum (zero gradient) at the two positions, i.e., at the top and the bridge site.

The nonlinear least squares fitting is a nontrivial enterprise, and unless special care is taken the result will often be poor. Therefore, the following important points have to be noted.

- The analytical expression should be physically meaningful, as simple as possible, and asymptotically correct.
- Start the minimization with the few most important parameters and with weights on the most important data.
- Add parameters one at a time and perform a new minimization, for example, by means of the weights turning on more and more details.

Table 4.4 gives the parameter values obtained by fitting to the top-site data at a fixed CO-Pt distance of 1.85Å. Although the fit is not extremely accurate, it gives the correct physical behaviour (see Fig. 4.5). Thus we see a bond stretch and a decrease in bond energy when the CO molecule approaches the surface.

It should be noted that dynamical calculations always include some averaging over phase space, and hence it is important to have a fit which is globally correct rather than to have one which is good locally. Thus a surface which is inadequate from a "spectroscopic" point of view can therefore be applicable from a dynamical point of view. The true surface will of course be able to describe more than one set of dynamical quantities. As the computer speed increases, it is possible in the nonlinear least squares fitting technique to incorporate other dynamical quantites which are not directly obtained from the potential but rather indirectly, by performing a dynamical calculation in the least squares fitting loop. In this manner several different types of experimental data can be fitted directly (see

TABLE 4.4 Least Squares Fitting to Top-Site CO-Pt(111) Parameters

Parameter	Value	Quantity	Fitted value
D_C	661.70 kJ/mol	ω_4	61 cm^{-1}
E_C	6.63 kJ/mol	ω_3	301 cm^{-1}
β_C	3.00 Å^{-1}	ω_2	345 cm^{-1}
D_O	0.022 kJ/mol	ω_1	2045 cm^{-1}
E_O	0.0024 kJ/mol	V_{min}	−127 kJ/mol
β_O	0.261Å^{-1}		
B_C	−0.0474 kJÅ^{-1}/mol		
B_O	0.0008 kJÅ^{-1}/mol		

The parameters C_C and C_O of eq. (4.42) are set to zero.

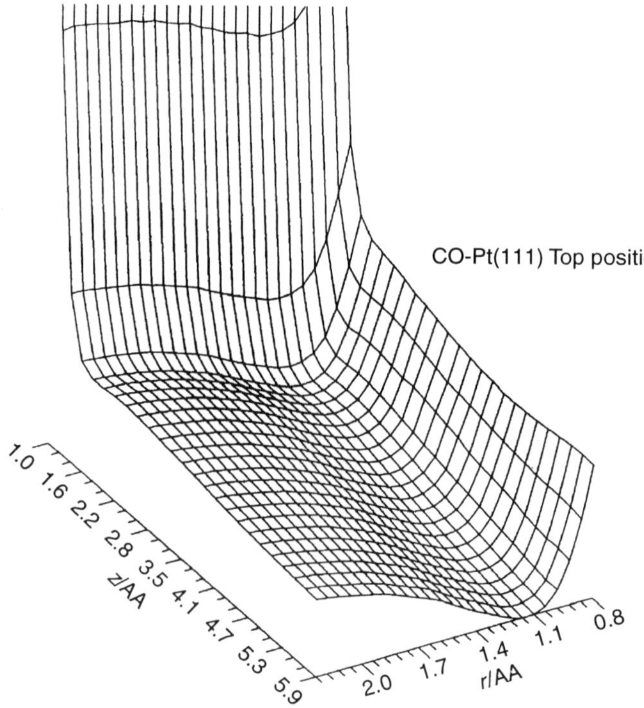

FIGURE 4.5 The CO-Pt(111) potential obtained by fitting a site-site expression to the spectroscopic frequencies and binding energy to the top-site.

Fig. 4.6). The standard way of improving a potential is to make the fitting to a set of ab initio points or perhaps to data such as vibrational frequencies, which are easily calculated from the potential. Then this "quess" is used in a dynamical calculation and, for instance, the barrier is modified to bring agreement with experimental data on sticking probabilities. However, the direct dynamical fitting may be feasible instead of this "trial-and-error"methodology. Certainly this would be the case if the dissociative sticking probability was calculated by transition state theory or any other simple dynamical method which would allow a fast and reliable calculation of it.

4.2.1 Multisurface Representation

Experimental beam data on some molecule-surface systems [149] show that the probability for dissociative sticking approaches a constant value (sometimes as small as 0.01–0.05) for energies well above the barrier. A possible and plausible explanation of this observation could be that a nonadiabatic transition from the lower adiabatic surface (see Fig. 4.7) to the upper surface, which is nonreactive, takes place. In order to incorporate such an effect in the dynamical calculations, we need to consider two configurations, AB + surface, AB^- + surface, or perhaps

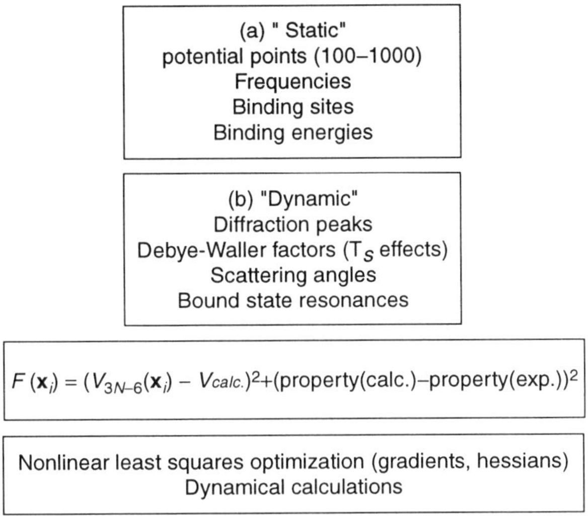

FIGURE 4.6 Potential energy surfaces for molecule-surface interaction can be obrtained by nonlinear least squares fitting to "static" and "dynamic" properties. The dynamical properties are obtained by doing a scattering calculation in each minimization step of the function $F(\mathbf{X}_i)$.

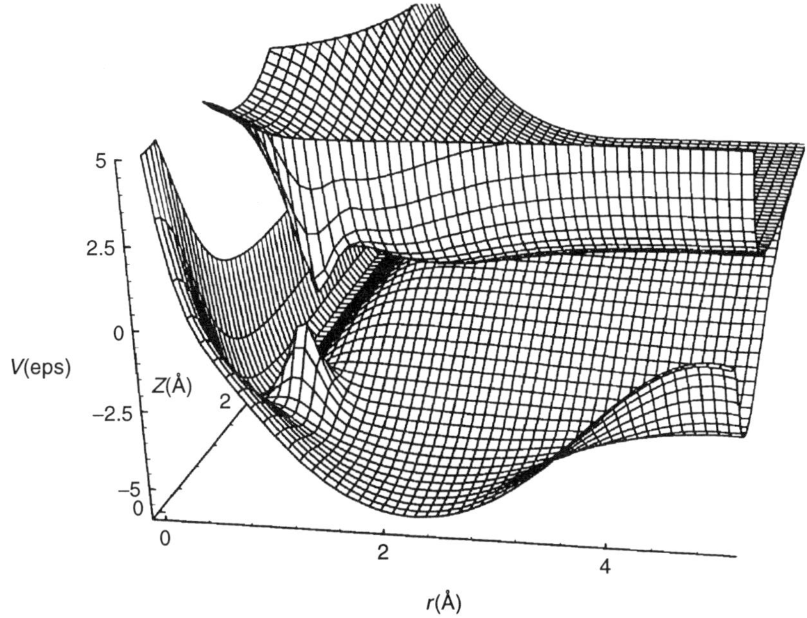

FIGURE 4.7 Molecule surface scattering often involves dynamical coupling between the two lower adiabatic potential energy surfaces. Nonadiabatic coupling between the two surfaces is large near the barrier for dissociation. Energy units eps = 100 kJ/mol.

a fully dissociative, $A + B +$ surface configuration. Let the diabatic potential energy surfaces associated with the two configurations be V_{11} and V_{22} respectively. A mixing term V_{12} is included to mix the configurations. We notice that the multisurface representation attempts to include the electron-transfer directly as a dynamical process rather than implicitly through the construction of the lower adiabatic surface. Simple models which deal with this problem are due to Gadzuk [154].

In the diabatic picture, the transition is caused by a coupling term $V_{12}(z, r)$, and hence the time-dependent Schrödinger equation (TDSE) which has to be solved involves two diabatic wavefunctions

$$i\hbar\frac{\partial}{\partial t}\begin{bmatrix}\Psi_1(z, r, t)\\ \Psi_2(z, r, t)\end{bmatrix} = \begin{bmatrix}V_{11}(z, r) & V_{12}(z, r)\\ V_{21}(z, r) & V_{22}(z, r)\end{bmatrix}\begin{bmatrix}\Psi_1(z, r, t)\\ \Psi_2(z, r, t)\end{bmatrix} \tag{4.44}$$

where $\Psi_i(z, r, t)$ is the diabatic wavefunction for surface i. In the case of, for instance, nitrogen-surface interaction, the diabatic potential V_{11} used to describe the molecule surface interaction, has been approximated as

$$V_{11}(z, r) = V_{N_2-M}(z, r) + V_{N_2}(r) \tag{4.45}$$

where the first term is the metal-molecule interaction and the last the gas-phase intramolecular potential. The diabatic V_{22} potential describe the atom or dissociated molecule-metal interaction

$$V_{22}(z, r) = V_{N-M}(z, r) + V_{N-N}(r) \tag{4.46}$$

where the last term in case of complete dissociation on surface 2 could be absent. Finally the diabatic coupling term is often represented by a Gaussian interaction in both coordinates

$$V_{12}(z, r) = C\exp(-\gamma_r(r - r^*)^2)\exp(-\gamma_z(z - z^*)^2) \tag{4.47}$$

In the simplest case we include only the dependence with z and r, but more realistic calculations try to incorporate the phonon interaction by representing the potential as a sum of pairwase atom-atom interactions [155]. The TDSE for the two diabatic surfaces can be solved using "grid-methods" [156]. Here the wavefunction is represented on a grid in the quantum coordinates, i.e., those appearing in eq. (4.44). The kinetic energy terms are evaluated using the FFT (fast fourier transform) technique described in section 10.3, and the time-propagation performed using, for example, the split operator technique [156,169].

Recent calculations have shown that the experimental data can be explained by incorporating the non-adiabatic coupling term V_{12} [170]. For the O_2–Ag system it is believed that one needs to incorporate three surfaces: one connected with the O_2–Ag interaction, one for the O–$O^{\delta-}$–Ag interaction, and finally the O-Ag surface [175]. The three surfaces for oxygen distinguish between the gas-phase molecule, an adsorbed molecular, and a dissociated state [176].

Processes such as photo-induced desorption involves more than one potential energy surface. In this case the wavepacket representing, say, a diatomic molecule is initialized at the surface in a surface-bound state, and a light source is used to bring the molecule to an upper surface with a dissociative repulsive dependence on the molecule-surface bond.

Information on the upper adiabatic surface is at present incomplete. In principle it is possible to perform the electronic structure calculations for such situations using either standard methods or, perhaps more conveniently, methods where density functional theory and standard ab initio methods are combined in a hybrid theory [144]. However, calculations of potential energy surfaces for electronically excited states have not yet been carried out nearly as extensively as for the ground state potential surface.

4.3 THE BOC-MP OR UBI-QEP METHODS

The semi-empirical methods referred to as bond-order conserving Morse potential (BOC-MP) or the unity bond index quadratic exponential potential (UBI-QEP) method have been reviewed extensively by Shustorovich and Sellers [145]. The UBI-QEP method is, like some of the empirical methods mentioned above, also based on the use of atom-atom interaction potentials, and here again a Morse form is convenient

$$E(r) = Q(x^2(r) - 2x(r)) \tag{4.48}$$

where Q is the well depth and $x(r)$ represents the bond order

$$x(r) = \exp(-a(r - r_0)) \tag{4.49}$$

The many, body potential is constructed by adding two-body terms

$$E(\mathbf{r}) = \sum_i Q_i(x_i(r_i)^2 - 2x_i(r_i)) \tag{4.50}$$

however, with the restriction that the sum of the bond orders x_i add up to unity, i.e., $\sum_i x_i = 1$.

From experimental information on heat of adsorption, one can obtain estimates of the parameters Q_i, wheres the additional parameters a_i and r_{0i} have to be determined from, for instance, spectroscopic information, dynamical data, or both.

The scheme has been build on the following postulates [145]:

1. The two-body interactions depend upon the atom-atom distance r_i only.
2. The two-body potentials have a single minimum, and approach zero as $r_i \to \infty$. The Morse potential obviously fulfills this condition.
3. The two-body potential is expressed as a polynomium in an exponential variable x_i.

4. The many-body potential is obtained by summing over nearest-neighbor pairwise interactions obeying the unity bond conservation law.

More specifically we would for an AB-surface potential have

$$E = Q_A(x_A^2 - 2x_A) + Q_B(x_B^2 - 2x_B) + D_{AB}(x_{AB}^2 - 2x_{AB}) \tag{4.51}$$

where D_{AB} is the Morse parameter for the diatomic molecule. The UBI constraint is $x_A + x_B + x_{AB} = 1$.

The UBI-QEP scheme is fairly simple to use, both for diatom and polyatom surface interactions. It is an attempt to include a large body of experimental data in the same theoretical framework. It has mainly been used to predict activation barriers for reactions and diffusion processes. However a few molecular dynamics studies using the UBI potentials for dissociative reactions on metal surfaces were reported in ref. [145]. However, in order to treat evenhandedly the reactant and product state, a modified potential function containing more parameters is suggested (see Appendix G of ref. [145]).

It has been argued that the metals do not act as a collection of atoms, but rather as a uniform entity. Thus various properties of the metal, such as elastic constants, are not accurately described by pair potentials [146]. If that is so, a model which assumes a pairwise atom-atom interaction will perhaps have to be replaced with one which incorporates the influence of electron density from a distributed electron gas rather than from localized ones. Such theories can furthermore be based on the Hohenberg Kohn theorem [211], namely that the energy is a functional of the electron density. Thus several approximate theories are based on the electron gas concepts, for example, the embedded atom and diatom methods, or the effective medium approach [210].

4.4 THE EDIM POTENTIAL

The embedded diatomics in molecules (EDIM) method is an extension of the embedded atom method (EAM) [146] to molecules using the diatomics in molecules (DIM) approach. The DIM method is based on a valence bond approach to electronically structure calculations. It constructs the polyatomic potential surfaces from the atom and diatom fragment ground and electronic excited states. Since these states are often well known either from experiment or calculations, the input to the scheme is often readily available. By construction, the potential obtained by the theory approaches the correct limit when the molecule is split into its various fragments [141].

In the EAM method the energy of N atoms is expressed as

$$V(\mathbf{R}) = \sum_i F_i(\rho_i(R)) + \frac{1}{2}\sum_{i \neq j} \phi_{ij}(R_{ij}) \tag{4.52}$$

where $\mathbf{R}$ is a collection of coordinate vectors for the N atoms, i.e.

$$\mathbf{R} = \{\mathbf{r}\}_{i=1}^{N} \tag{4.53}$$

The distance between two atoms is $R_{ij} = |\mathbf{r}_i - \mathbf{r}_j|$ and ϕ_{ij} is a short-range "effective" Coulomb potential

$$\phi_{ij}(R) = CZ_i(R)Z_j(R)/R \tag{4.54}$$

where $Z_i(R)$ is the effective charge on nucleus i at the distance R. The function $F_i(\rho)$ is the energy associated with the embedding of atom i in an electron gas with density ρ_i. The electron density a given atom i feels is obtained by summing over the density created at the location of atom i by all the remaining atoms, i.e.,

$$\rho_i(\mathbf{R}) = \sum_{j \neq i} \rho_j^a(R_{ij}) \tag{4.55}$$

The embedding function $F_i(\rho)$ is "universal", i.e., it does not depend upon the source which creates the electron density. The atomic densities are usually calculated from the tabulated Hartree-Fock wavefunctions [142] and for transition metals taken as a linear combination of the density created by the valence s- and d-electrons, i.e.,

$$\rho^a(R) = n_s\rho_s(R) + n_d\rho_d(R) \tag{4.56}$$

where n_s and n_d are the number of s and d electrons in the valence shell, respectively.

Sometimes the total number of electrons $n_s + n_d$ is fixed and n_s fitted to the properties of the metal [148]. The atomic densities are given in the tables of Clementi et al. [142] and McLean and McLean [143]. They are represented by the expression

$$\rho_{s,d}(r) = \frac{1}{4\pi}\left|\sum_i C_i R_i(r)\right|^2 \tag{4.57}$$

where $R_i(r)$ is:

$$R_i(r) = \frac{(2\zeta_i)^{n_i+(1/2)}}{\sqrt{(2n_i)!}} r^{n_i-1}\exp(-\zeta_i r) \tag{4.58}$$

where the coefficients C_i and ζ_i are available from the Clementi/McLean tables.

The embedding function $F(\rho)$ is as mentioned supposedly universal but has been determined for various metals [177] and also given an analytical form (see, e.g., Truong et al. [209]). The effective charge $Z(R)$ is expressed as [177]

$$Z(R) = Z_0(1 + \beta R^{\nu})\exp(-\alpha R) \tag{4.59}$$

where Z_0, β, ν, and α are parameters (see Table 4.5). We notice that the embedding energy $F(\rho)$ is negative, whereas the last term in (4.52) is a Coulomb repulsion. Modified expressions with more parameters have been suggested in refs. [147,177]. These expressions can better reproduce phonon dispersion curves [147].

In order to extend the EAM approach to molecules, it has been suggested that the DIM method should be used and "pseudo-atoms" representing the metal introduced, proceeding as in the ordinary DIM scheme. As mentioned above, one then introduces for a diatomic molecule interacting with a metal altogether four orbitals: two non metal (a and b) and two metal orbitals (c and d) where c and d are treated as an "extended" electron source. In the DIM method the lowest adiabatic surface W_{11} is obtained by solving the secular equation:

$$|\mathbf{H} - \mathbf{S}W| = 0 \tag{4.60}$$

where the elements of the $\mathbf{H}$ matrix are constructed using the Rumer diagrams [174] (see also Appendix B) and $\mathbf{S}$ is an overlap matrix. Thus with four orbitals we have:

$$H_{11} = 4(Q + J_{ab} + J_{cd}) - 2(J_{ac} + J_{ad} + J_{bc} + J_{bd}) \tag{4.61}$$

$$H_{22} = 4(Q + J_{ad} + J_{bc}) - 2(J_{ab} + J_{ac} + J_{bd} + J_{cd}) \tag{4.62}$$

and

$$H_{12} = -2(Q + J_{ab} + J_{ad} + J_{bc} + J_{cd} - 2J_{ac} - 2J_{bd}) \tag{4.63}$$

where

$$Q = Q_{ab} + Q_{ac} + Q_{ad} + Q_{bc} + Q_{bd} + Q_{cd} \tag{4.64}$$

Q and J are Coulomb and exchange integrals for the diatomic fragments. As in the ordinary LEPS theory these will not be used directly—only when constructing

TABLE 4.5 Embedding Parameters for Various Metals

Metal	Cu	Ag	Au	Ni	Pd	Pt
Z_0	11.0	11.0	11.0	10.0	10.0	10.0
α Å^{-1}	1.7227	2.1395	1.4475	1.8633	1.2950	1.2663
β	0.1609	1.3529	0.1269	0.8957	0.0595	0.1305
ν	2	2	2	1	1	1
n_s	1.000	1.6760	1.0809	1.5166	0.8478	1.0571
a_0 Å	3.615	4.09	4.08	3.52	3.89	3.92

Data from [177].

the functional relationship between the fragment states and the many-body interaction potential. Thus we have:

$$Q_{ij} = \tfrac{1}{2}(E^s_{ij} + E^t_{ij}) \tag{4.65}$$

$$J_{ij} = \tfrac{1}{2}(E^s_{ij} - E^t_{ij}) \tag{4.66}$$

where s, t denote singlet, triple nonding and antibonding energies.

The metal is now treated as an extended electron source, i.e., the Coulomb and exchange terms for the *c-d* interaction as well as the redundent *a-d* and *b-c* terms are removed. Thus we get the following simplifications:

$$H_{11} = 4(Q + J_{ab}) - 2(J_{ac} + J_{bd}) \tag{4.67}$$

$$H_{22} = 4(Q + J_{ad} + J_{bc}) - 2J_{ab} \tag{4.68}$$

$$H_{12} = -2(Q + J_{ab} + J_{ad} + J_{bc}) \tag{4.69}$$

$$Q = Q_{ab} + Q_{ac} + Q_{bd} \tag{4.70}$$

The overlap matrix elements are $S_{11} = S_{22} = 4$ and $S_{12} = -2$. Thus if we know the various exchange and Coulomb integrals we can evaluate the lower adiabatic surface. From equations (4.65), and (4.66) above, we see that all we need to do is to relate these integrals to the diatomic fragment singlet and triplet curves. Thus for the two gas-phase atoms a and b we have

$$E^s_{ab} = D_{ab}(1 - \exp(-\alpha_{ab}(R_{ab} - R^{eq}_{ab})))^2 - D_{ab} \tag{4.71}$$

$$E^t_{ab} = \frac{1}{2}\frac{1 - S_{ab}}{1 + S_{ab}}(D_{ab}(1 + \exp(-\alpha_{ab}(R_{ab} - R^{eq}_{ab})))^2 - D_{ab}) \tag{4.72}$$

i.e., the Coulomb and exchange integrals can be found from the Morse and anti-Morse curves for the diatomic fragment. The singlet interaction between the gas-phase atom and the metal is modeled by the EAM method, i.e.,

$$E^s_{aM} = F_a(\overline{\rho}_a(\mathbf{R})) + \sum_{\alpha=1}^{N} \phi_{a\alpha}(R_{a\alpha}) \tag{4.73}$$

where F_a is the embedding function for the atom a. It is sometimes expressed in a continued fraction form

$$F_a(\rho) = \delta_1\rho/(1 + \delta_2\rho/(1 + \delta_3\rho/(1 + \delta_4\rho))) \tag{4.74}$$

with 4 parameters $(\delta_1 - \delta_4)$. In eq. (4.73) $\overline{\rho}_a$ is the electron density created by the metal atoms at the position of atom a

$$\overline{\rho}_a(\mathbf{R}) = \sum_{\alpha=1}^{N} \rho^a_\alpha(R_{\alpha a}) \tag{4.75}$$

where the sum runs over the metal atoms and ρ^a is an atom density obtained as described above. The DIM idea is now extended to metals in a rather ad hoc fashion by introducing the "triplet" energies as

$$E^t_{aM} = -F_a(\overline{\rho}_a(\mathbf{R})) + \sum_{\alpha=1}^{N} \phi_{a\alpha}(R_{a\alpha}) \tag{4.76}$$

The energy defined in this manner is then purely repulsive and is sometimes modefied by multiplying with an empirical function instead of the "Sato factor" $(1 - S_{ab})/(1 + S_{ab})$ in front of the gas-phase triplet energy in eq. (4.72). In eq. (4.73) $\phi_{a\alpha}$ is a screened Coulomb potential as given by eq. (4.54) and hence $Z_a(R)$ is expressed as

$$Z_a(R) = Z_0^a(1 + \beta_a R^{\nu_a}) \exp(-\alpha_a R) \tag{4.77}$$

Thus when fitting the EDIM potential we have the parameters δ_i in the embedding function and for each different atom in the gas-phase the parameters $Z_0^a, \beta_a, \nu_a, \alpha_a$. Thus for a homonuclear molecule this scheme would involve 7 parameters [179]. The EDIM method is a semi-empirical method which gives the correct asymptotic potential curves. It is based on ideas similar to those in the EAM (embedded atom) and effective medium theories [210]. It contains enough parameters to fit ab initio data at a given site. The method gives a global fit but may not be able to reproduce all details locally. In a sense, the EDIM potential is the surface equivalent of the gas-phase LEPS potential. A normal approach in gas-phase chemistry is to start with the LEPS surface, and then to add corrections, i.e., three- and four-body terms. In surface chemistry a similar methodology could be build around the EDIM potential.

4.5 DENSITY FUNCTIONAL THEORY

In order to obtain the potential energy surface, one needs of course to solve an electronic structure problem. Several methods have been used, ranging from methods treating the electrons of the metal as a free electron gas to "traditional" ab initio calculations. One of the simplest models treats the solid as a jellium, i.e., a uniform elecron gas on a positive background. The hamiltonian for the electrons is then just the kinetic energy $-(\hbar^2/2m_i)\nabla^2$. For $z > 0$ the one-electron wavefunctions are approximated by

$$\psi_i = C_{k_x k_y k_z} \exp(i(k_x x + k_y y)) \exp(-z\sqrt{2m_i E/\hbar^2 - k_z^2}) \tag{4.78}$$

i.e., a wavefunction which has an exponential tail in the vacuum. Methods for solving the Kohn-Sham equations (see Chapter 10) for adatom-jellium systems have been devised by Lang and Williams [182].

In the effective medium theory, the electronic interaction between an atom and the solid is replaced by that of the atom and a homogeneous electron gas. The approach is based upon density functional theory, but since the solid is not really behaving as a homogeneous electron gas, gradient corrections to the theory have been devised [210]. Simple model hamiltonians have also been used to study chemisorption of atoms and molecules at metals. Here we mention the Anderson-Grimley-Newns model hamiltonian of the type

$$H = \varepsilon_a n_a + \sum_k \varepsilon_k n_k + V(t) \sum_{ak} (c_a^+ c_k + c_a c_k^+) \hbar\omega b^+ b + \Delta n_a (b^+ + b) \quad (4.79)$$

where n_a and n_k are number operators for the electron state of the molecule and the surface electrons, respectively. ε_a is the energy of the electronic level of the molecule. $V(t)$ is the potential for the interaction between the molecule and the surface electrons. Creation/anhilation operators for the vibrational mode of the molecule are b^+ and b, respectively, and ω the vibrational frequency. The last term describes electron/vibration coupling in the molecule with a coupling parameter Δ. Although perturbation theory has been used to solve the time-dependent Schrödinger equation

$$i\hbar \frac{\partial \Psi}{\partial t} = H(t)\Psi(t) \quad (4.80)$$

the above hamiltonian (4.79) is of the type which allows a solution of the TDSE to infinite order, using operator algebraic methods. We shall return to this problem in Chapter 11. However, despite the importance of the above-mentioned approximate theories, the development has more recently been in the direction of using the density functional theory itself rather than to apply various approximate versions of it.

The general theory for electron density, density functional theory (DFT) is due to Hohenberg and Kohn [211]. It was shown that the ground state energy of the many-particle system was related to the density distribution of the electrons. The proof of the theory was later extended and simplified by Levy [214]. Kohn and Sham [215] derived a system of self-consistent equations for the effective one-particle potential

$$v_{eff}(\mathbf{r}) = \phi(\mathbf{r}) + v_{xc}(\mathbf{r}) \quad (4.81)$$

where the first term is the electrostatic potential and the last the exchange-correlation contribution. This term is also a functional (but complicated) of the density. If approximate expressions can be found for this part the DFT theory is as simple to use as the Hartree theory but it yields much better results (see Table 4.6). The equation for the wavefunction is

$$\left[-\frac{\hbar^2}{2m_e} \nabla^2 + v_{eff}(\mathbf{r}) \right] \psi_i = \varepsilon_i \psi_i \quad (4.82)$$

TABLE 4.6 Comparison of Results for Bond Lengths r_e, Vibrational Frequency ω_e, and Atomization Energies E_d for Some Molecules

Molecule	Property	HF	LSD	GC	Experiment
H_2	r_e Å	0.730	0.781	0.748	0.741
	ω_e cm^{-1}	4646	4035	4373	4160
	D_e kJ/mol	318	321	432	432
CO	r_e Å	1.114	1.153	1.150	1.128
	ω_e cm^{-1}	2439	2105	2105	2170
	D_e kJ/mol	704	1080	1077	1072
F_2	r_e Å	1.345	1.407	1.434	1.417
	ω_e cm^{-1}	1245	1052	987	923
	D_e kJ/mol	−144	288	228	154

HF denotes Hartee Fock theory, LSD the local spin density with no correlation but exchange according to a free electron gas, and GC has exchange according to ref. [230] and correlation according to ref. [229]. Thus LSD refers to the simplest and GC to a more state-of-the-DFT theory. Data adapted from ref. [236].

and the density is

$$n(\mathbf{r}) = \sum_{i=1}^{N} \psi_i^* \psi_i \tag{4.83}$$

The equations must be solved iteratively since the density enters the expressions for v_{eff}. The free electron gas, the Fermi-Thomas, and the DFT theory will be treated in more detail in Chapters 9 and 10. The DFT theory has been used recently in a number of calculations of molecule-surface interaction (see Table 4.7). Some studies just involves the chemisorption to a small cluster of metal atoms [189], and it is questionable how well this represents the behavior of a bulk metal.

The DFT calculations provide only the lower adiabatic surface, and nonadiabatic effects are therefore not accounted for. But the calculations are feasible to perform and therefore in principle provide the necessary input to empirical models such as the EDIM model and they allow a determination of the parameters in these models. But it should be emphasized that a detailed understanding of surface catalytic effects involves more than the static picture provided by the DFT calculations. It involves the actual simulation of the dynamics, including the important dynamical aspects as mentioned in the Chapter 1. The information provided by the DFT calculations is, however, the first important step in the direction of achieving this goal. Although the DFT approach may fail to give accuracy better than about 0.1 eV for molecule-surface interaction potentials, it is hard at present to see any alternative methods which can routinely give sufficient information on the interaction potential for systems involving enough atoms

TABLE 4.7 Some Recent Density Functional Calculations on Molecule Surface Interaction

System	Reference
$H_2 + Cu(100)$	[178, 180, 184]
$H_2 + Cu(111)$	[181]
$H_2 + Mg(0001)$	[186]
$H_2 + Al(100)$	[183]
$H_2 + NiAl(110)$	[185]
$H_2 + Pt(111)$	[200]
$H_2 + W(100)$	[187]
$H_2 + Si(100)$	[188]
$NO + Pd(112)$	[190]
N_2 and $CO + Ru(0001)$	[191]
$H_2O + TiO_2$	[192]
O_2, N_2, CO, $NO + Cu(100)$	[193]

to be of value for simulating molecule-surface dynamics. Thus DFT methods are routinely used for calculating the ground state energy of more complicated systems, including not only surfaces but also, for instance, polyatomic molecules, clusters, zeolites, and so forth.

4.6 INTRAMOLECULAR POTENTIALS

For diatomic molecules, it is often sufficiently accurate to describe the interaction by using a Morse potential. This has also been explicitly assumed in some of the previous sections. However, near the dissociation limit additional accuracy can be obtained by using a functional form containing additional parameters. Such a function is the extended Rydberg (ER) function, given by

$$V = -D_e\left(1 + \sum_k a_k(r - r_e)^k\right)\exp(-\gamma(r - r_e)) \tag{4.84}$$

The parameters D_e, a_k, r_e, and γ are fitted to spectroscopic data.

For polyatomic molecules the situation is more complex, and detailed knowledge of the potential hypersurface is available only for few systems. Semi-empirical methods are therefore introduced. As in the construction of the LEPS surface one should in the semi-empirical method, use expressions which behave correctly in the asymptotic limit. Thus for the molecule ABC the potential should approach V_{AB} for $R_{BC} \to \infty$, V_{BC} for $R_{AB} \to \infty$, and V_{AC} for $R_{AB} \to \infty$. The LEPS function discussed above has this property. However, it is advantageous to have a recipe which uses the extensive information on force-field potentials which is available from spectroscopic measurements. For a molecule as CO_2, the parameters in the expression

$$
\begin{aligned}
W(\Delta R_1, \Delta R_2, \Delta\alpha) = {} & K_{11}(\Delta R_1^2 + \Delta R_2^2) + K_{12}\Delta R_1 \Delta R_2 + K_{33}\Delta\alpha^2 \\
& + K_{111}(\Delta R_1^3 + \Delta R_2^3) + K_{133}(\Delta R_1 + \Delta R_2)\Delta\alpha^2 \\
& + K_{122}(\Delta R_1 + \Delta R_2)\Delta R_1 \Delta R_2 + K_{1111}(\Delta R_1^4 + \Delta R_2^4) \\
& + K_{1112}(\Delta R_1^2 + \Delta R_2^2)\Delta R_1 \Delta R_2 + \cdots
\end{aligned} \tag{4.85}
$$

are obtainable from spectroscopy data. In eq. (4.85), $\Delta R_i = R_i - R_i^e$ (the equilibrium distance) and $\Delta\alpha = \alpha - \pi$. In order to obtain the correct asymptotic form of the intramolecular potential for the CO_2 system, we use

$$V(R_1 R_2 R_3) = V_{CO}(R_1) + V_{CO}(R_2) + V_{O_2}(R_3) + V_I(R_1 R_2 R_3) \tag{4.86}$$

where V_{CO} and V_{O_2} are the asymptotic Morse or ER potentials and V_I a three-body interaction term. For this term it has been suggested to introduce [194]

$$V_I = AP(\Delta R_1 \Delta R_2 \Delta R_3)\Pi_{i=1}^3(1 - \mathrm{th}(\gamma_i \Delta R_i)) \tag{4.87}$$

The switching functions (1 − th) makes the three-body interaction term vanish in the limit $\Delta R_i \to \infty$. The function $P(\ldots)$ is a polynomial in ΔR_i, i.e.,

$$
\begin{aligned}
P = 1 + \varepsilon_1 \Delta R_1 + \varepsilon_2 \Delta R_2 + \varepsilon_3 \Delta R_3 + \varepsilon_{11}\Delta R_1^2 + \\
\varepsilon_{22}\Delta R_2^2 + \varepsilon_{33}\Delta R_3^2 + \varepsilon_{12}\Delta R_1 \Delta R_2 + \cdots
\end{aligned} \tag{4.88}
$$

The coefficients ε_i, ε_{ij}, and so on can now be found by expanding $\mathrm{th}(\gamma_i \Delta R_i)$ in a Taylor series and comparing eq. (4.87) and eq. (4.85). The scheme is rather heavy but straightforward and the appropriate algebraic transformations have appeared in the literature [194]. Note that it is only the expansion parameters A, ε_i, and so on which are determined in this manner. The parameters γ_i should be determined in such a way that $1 - \mathrm{th}(\gamma_i \Delta R_i) \sim \exp(-2\gamma_i \Delta R_i)$ vanishes as fast as the Morse or ER potential in the limit $\Delta R_i \to \infty$.

A semi-empirical scheme of the type described here has been used to construct a potential for the interaction between CO_2 and a Pt(111) surface. In ref. [196] this potential was used to study oxidation of CO on a platinum surface.

4.7 CRYSTAL ENERGIES

In order to obtain properties of the isolated crystal as, for instance, the phonon spectrum, we need to calculate the crystal or lattice energy. We have already in the section about the EDIM potential given an expression obtained with the EAM (embedded atom method). Other methods expresses the total potential as a sum of pairwise Lennard-Jones, exp-6, Morse, or even harmonic potentials

$$V_c = \frac{1}{2} \sum_{ij} V_{ij}(r_{ij}) \tag{4.89}$$

where $V_{ij}(r)$ is the pair potential between the atoms i and j ($i \neq j$). By comparing with, for example, experimental data for the phonon spectrum, the parameters of the pair potential can be determined.

If the solid consists of ions the semi-empirical potential often used is the so-called Born-Mayer potential

$$V(r_{ij}) = A \exp(-ar_{ij}) + \frac{z_i z_j e^2}{r_{ij}} \tag{4.90}$$

where z_i and z_j are the ionic charges and r_{ij} the distance. between the ions. The sum over the Coulomb term in eq. (4.89) is problematic, but can be carried out by using a method known as Ewald summation. Let the positions of the positive ions be given by the vector

$$\mathbf{R} = n_1 \mathbf{a}_1 + n_2 \mathbf{a}_2 + n_3 \mathbf{a}_3 \tag{4.91}$$

where n_i are integers and $\mathbf{a}_i$ lattice vectors. The contribution to the Coulomb term from the interaction between a positive ion positioned at the origin and the remaining positive ions is then

$$\sum_{n_1 n_2 n_3} \frac{e^2}{|\mathbf{R}|} \tag{4.92}$$

where for simplicity of the notation we have set $z_i = +1$ for positive and $z_j = -1$ for negative ions. The negative ions are located at $\mathbf{R} + \mathbf{x}$ where $\mathbf{x}$ can be expressed as

$$\mathbf{x} = i_1 \mathbf{a}_1 + i_2 \mathbf{a}_2 + i_3 \mathbf{a}_3 \tag{4.93}$$

where $i_j j = 1, 2$, and 3 are fixed integers, which defines the location of a negative ion relative to a positive. The Coulomb energy from the interaction between the positive ion at (0,0,0) and all negative ions is

$$-\sum_{n_1 n_2 n_3} \frac{e^2}{|\mathbf{R} + \mathbf{x}|} \tag{4.94}$$

With N positive ions we then obtain the total lattice energy as

$$V_c = \frac{Ne^2}{2} \sum_{n_1 n_2 n_3} \left(\frac{1}{|\mathbf{R}|} - \frac{1}{|\mathbf{R} + \mathbf{x}|} \right) \tag{4.95}$$

where we have divided by two in order not to count the interaction twice. We note that the integers n_i should not all be zero. We also notice that the same expression would have been obtained if the position of a negative ion had been used to define the origin. We now use the identity

$$\frac{1}{|\mathbf{R}+\mathbf{x}|} = \frac{2}{\sqrt{\pi}} \int_0^{\infty} dz \exp(-|\mathbf{R}+\mathbf{x}|^2 z^2) \tag{4.96}$$

and utilizing the lattice periodicity we can write the function

$$F(z, \mathbf{x}) = \frac{2}{\sqrt{\pi}} \sum_{n_1 n_2 n_3} \exp(-|\mathbf{R}+\mathbf{x}|^2 z^2) \tag{4.97}$$

in terms of the reciprocal lattice vectors $\mathbf{G}$ (see Chapter 2) as

$$F(z, \mathbf{x}) = \sum_G F_G(z) \exp(i\mathbf{G}\cdot\mathbf{x}) \tag{4.98}$$

and hence the Fourier components $F_G(z)$ can be obtained as

$$F_G(z) = \frac{1}{V} \sum_{n_1 n_2 n_3} \frac{2}{\sqrt{\pi}} \int d\mathbf{x} \exp(-|\mathbf{R}+\mathbf{x}|^2 z^2) \exp(-i\mathbf{G}\cdot\mathbf{x}) \tag{4.99}$$

where V is the volume of the crystal. Using now the fact that $\exp(-i\mathbf{G}\cdot\mathbf{R}) = 1$ we get

$$F_G = \frac{2}{V\sqrt{\pi}} \sum_{n_1 n_2 n_3} \int d\mathbf{q} \exp(-q^2 z^2 - i\mathbf{G}\cdot\mathbf{q}) \tag{4.100}$$

where $\mathbf{q} = \mathbf{R}+\mathbf{x}$. The integral over $\mathbf{q}$ can be evaluated to give

$$F_G(z) = \frac{2\pi}{v_c} \frac{1}{z^3} \exp(-G^2/4z^2) \tag{4.101}$$

where $v_c = V/N$ is the volume of a unit cell (see Table 2.4). Thus we have

$$F(z, \mathbf{x}) = \frac{2\pi}{v_c} \sum_G \int_0^{\infty} dz \frac{1}{z^3} \exp(-G^2/4z^2) \exp(i\mathbf{G}\cdot\mathbf{x}) \tag{4.102}$$

where the integral over z can be carried out. If the formula is used also for $\mathbf{x} = \mathbf{0}$, we get

$$\sum_{n_1 n_2 n_3} \left(\frac{1}{|\mathbf{R}|} - \frac{1}{|\mathbf{R}+\mathbf{x}|} \right) = \frac{4\pi}{v_c} \sum_G \frac{1}{G^2} (1 - \exp(-i\mathbf{G}\cdot\mathbf{x})) \tag{4.103}$$

where $\mathbf{G} = m_1\mathbf{b}_1 + m_2\mathbf{b}_2 + m_3\mathbf{b}_3$ and using that $\mathbf{G}\cdot\mathbf{x} = 2\pi(m_1 i_1 + m_2 i_2 + m_3 i_3)$ we finally get for an SC, BCC, and an FCC lattice (see Table 2.4)

$$V_c = \frac{a^2 e^2}{\pi v_c} \sum_{m_1 m_2 m_3} \frac{1}{m_1^2 + m_2^2 + m_3^2}(1 - \cos(2\pi(m_1 i_1 + m_2 i_2 + m_3 i_3))) \quad (4.104)$$

We notice that there is no singularity at $m_j = 0$ and that the total energy, the Madelung energy, can be written as

$$V_c = \frac{N\alpha_M e^2}{r_0} \quad (4.105)$$

where $r_0 = v_c/a^2$ is proportional to the lattice constant and the Madelung constant defined as

$$\alpha_M = \frac{1}{2\pi} \sum_{m_1 m_2 m_3} \frac{1 - \cos(2\pi(m_1 i_1 + m_2 i_2 + m_3 i_3))}{m_1^2 + m_2^2 + m_3^2} \quad (4.106)$$

is a dimensionless constant characteristic for the given crystal structure. The above way of treating coulombic interactions in a periodic environment will also be useful in density functional theory (see, e.g., [195]). Density functional theory is briefly reviewed in Chapter 10.

5

QUANTUM TREATMENT OF ATOM/MOLECULE-SURFACE SCATTERING

If one neglects the phonon coupling and focuses on the z-dependence and the corrugation as obtained from a perfect lattice, the interaction potential between an atom and the solid depends only upon the position coordinates x, y, and z of the atom. Thus the potential $V(x, y, z)$ can be expressed in terms of a Fourier expansion

$$V(x, y, z) = \sum_{G} \exp(i\mathbf{G} \cdot \mathbf{R})V_G(z)) \tag{5.1}$$

where $\mathbf{R} = (x, y)$ and $\mathbf{G} = n\mathbf{G}_1 + m\mathbf{G}_2$. Here $\mathbf{G}_i$ are the reciprocal lattice vectors and n and m integers labeling the diffraction channels.

Scattering to the diffraction channel $(n, m) = (0, 0)$ is called specular scattering. The Schrödinger equation for atom-surface scattering is

$$-\frac{\hbar^2}{2m}\nabla_r^2\psi(\mathbf{r}) + V(\mathbf{r})\psi(\mathbf{r}) = E\psi(\mathbf{r}) \tag{5.2}$$

where m is the mass of the atom, $E = (\hbar^2k^2/2m)$ and $\mathbf{k}$ the wavevector. With the potential (5.1) we have solutions of the form

$$\Psi_K(\mathbf{r}) = \sum_{G=(n,m)} \exp(i(\mathbf{K} + \mathbf{G}) \cdot \mathbf{R})\Psi_G(z) \tag{5.3}$$

where $\mathbf{K}$ is the projection of the wavevector on the surface, i.e., if θ is the incident angle (the angle to the surface normal) we have $K_x = k \sin\theta \cos\phi$, $K_y = k \sin\theta \sin\phi$, and

$$K = k \sin\theta \tag{5.4}$$

Inserting (5.3) in (5.2) and using that $\nabla_r^2 = (\partial^2/\partial x^2) + (\partial^2/\partial y^2) + (\partial^2/\partial z^2)$ we get

$$-\frac{d^2\Psi_G(z)}{dz^2} + \frac{2m}{\hbar^2}\sum_{G'} V_{G-G'}(z)\Psi_{G'}(z) = k_G^2\Psi_G(z) \tag{5.5}$$

where

$$k_G^2 = k^2 - (\mathbf{K}+\mathbf{G})^2 \tag{5.6}$$

Thus we have obtained a set of coupled second-order differential equations in the diffraction channels. The equations (5.5) may be solved efficiently by a number of methods, such as the log-derivative method, in which $\Phi_G = (d \ln \Psi_G/dz) = (d\Psi_G/dz)\Psi_G^{-1}$ is propagated instead of Ψ_G [126]. Thus substitution transforms the second-order differential equation to a first-order nonlinear Ricatti equation.

The so-called closed channels have $k_G^2 < 0$ and the scattered or open channels have $k_G^2 > 0$. The proper solution to the coupled-channel equation satisfies the boundary conditions

$$\Psi_{00}(z) = \exp(-ik_0 z) + A_0 \exp(ik_0 z) \tag{5.7}$$

where the first term is an incoming wave and the second the reflected specular wave with amplitude A_0 and wavenumber k_0. Note that

$$k_0^2 = k^2 - K^2 = k_z^2 \tag{5.8}$$

and $k_z = k\cos\theta$. For the diffraction channels we have

$$\Psi_G(z) = A_G \exp(ik_G z) \tag{5.9}$$

The intensity in the diffraction channel is defined by $|S_{0G}|^2$, where S_{0G} is an element of the scattering S-matrix. The probability for scattering in diffraction channel G, $|S_{0G}|^2$ is obtained as the ratio between the outgoing and incoming fluxes, i.e.,

$$|S_{0G}|^2 = \frac{\text{outgoing flux}}{\text{incoming flux}} = \frac{k_G}{k_0}|A_G|^2 \tag{5.10}$$

where we have used the fact that the quantum flux is defined by $(i\hbar/m)(\Psi^*(d/dz)\Psi - \Psi(d/dz)\Psi^*)$. A_G is obtained from the solution to eq. (5.5), subject to the above-mentioned boundary conditions.

The extension to diatom-surface scattering is in principle straightforward, but the number of coupled equations increases since rotational and vibrational states of the diatom also need to be included. Thus the expansion of the wavefunction eq. (5.3) includes additional quantum numbers. For a diatomic molecule we have the additional quantum numbers v, j, m_j for the vibration, rotational, and rotational projection quantum states, respectively. This rather pedestrian approach leads to the necessity of solving a large number of coupled second-order differential equations and this is convenient or even possible only for low energies involving light molecules, where phonon excitation can be neglected and where the number of energetically open vibrational/rotational states is small [128]. However, the well-known decoupling techniques, known from gas-phase dynamics as the infinite order sudden (IOS) [125] and coupled states (CS) [129] methods, can of course also be used in molecule surface scattering [127]. In these decoupling methods the rotational or the centrifugal operators are replaced with effective values of the rotational and orbital angular momenta. As a consequence, the coupling scheme is drastically reduced. This is necessary since the number of coupled second-order differential equations which can be solved on computers with today's, teraflop (10^{12} operations per second) performance is about 10,000. If all the quantum indices are maintained, this number is so small that "exact" or close-coupling methods can be used only on a few systems at low energies.

To include a large number of phonon states in a full quantum description is in principle also straightforward, although from a numerical point of view rather cumbersome [158]. It is, therefore, more convenient to deal with these processes in different approximate fashions, to be described in later chapters.

At higher energies where many quantum states are energetically accessible and where also chemical reaction (dissociation) can take place, it is advantageous (see discussion below) to turn to a time-dependent description, where the time-dependent Schrödinger equation (TDSE) is solved. That is, instead of eq. (5.2) we would have

$$i\hbar\frac{\partial}{\partial t}\psi(x, y, z, t) = -\frac{\hbar^2}{2m}\nabla_r^2\psi(x, y, z, t) + V(x, y, z)\psi(x, y, z, t) \tag{5.11}$$

For a diatomic molecule additional variables r, the bond distance, and θ, ϕ, the orientation of the molecule, enter. Hence the diffractive wave function will be

$$\Psi_G(z, r, \theta, \phi, t) \tag{5.12}$$

This wave function can be expanded in a basis set consisting of spherical harmonics Y_{jm}, as

$$\Psi_G(z, r, \theta, \phi, t) = \sum_{jm_j} C_{G,j,m_j}(z, r, t)Y_{jm_j}(\theta, \phi) \tag{5.13}$$

where $G = (n, m)$ labels the diffraction channels. The coordinates z and r are now conveniently represented on a two-dimensional grid (see Fig. 5.1). The wavepackets $C_{Gjm_j}(z, r, t)$ are initialized at a large value of z and propagated as coupled wavepackets using a fast Fourier transform (FFT) method (see Chapter 10) for evaluating the effect of the kinetic energy operators in r and

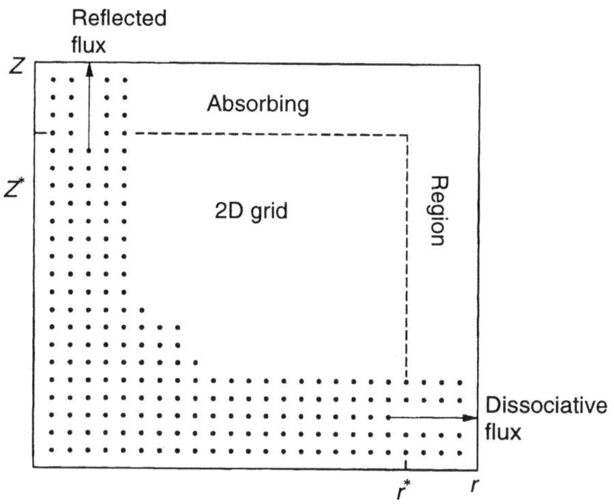

FIGURE 5.1 The grid used to propagate a 2D wavepacket. The reflected and dissociative fluxes are recorded at Z^* and r^* before absorbing the wavefunction with a negative imaginary potential in the absorbing region.

z on the wavefunction. For the time propagation (i.e., the solution of eq. (5.11)), a number of efficient methods exist (see, e.g., refs. [124,156]). The advantage of treating the r and z coordinates as grid-variables has to do with the possibility of dissociation and tunneling which takes place in these variables. Thus, part of the wavepacket will dissociate (go to areas with large r-values) while another part will be scattered back.

The back-scattered part can be analyzed by projecting on the appropriate asymptotic (large z values) eigenstates, i.e., vibrational/rotational states of the molecule times an outgoing plane wave in the z direction. Thus the amplitude for scattering in a given molecular state and diffraction channel is obtained as

$$A_G^{vjm_j} = \lim_{t\to\infty} \frac{1}{c_{k_0}\sqrt{2\pi}} \int dz \int dr \exp(-ik_G z)\phi_{vj}(r) C_{G,j,m_j}(z, r, t) \tag{5.14}$$

where c_{k_0} is the initial weight of the plane wave component k_0.

Thus, by initializing using a Gaussian wavepacket in z we can cover an energy range in one propagation. This has to do with the fact that a Gaussian wavepacket can be resolved into a distribution of plane waves, i.e.,

$$\Phi_{GWP}(z, t_0) = \sum_{k_0} c_{k_0} \exp(-ik_0 z) \tag{5.15}$$

where $\Phi_{GWP}(z, t_0)$ is the initial Gaussian wavepacket

$$\Phi_{GWP}(z, t_0) = (2\mathrm{Im}\alpha_0/\pi)^{1/4} \exp(i(\alpha_0(z - z_0)^2 - ik(z - z_0))) \tag{5.16}$$

where α_0 is a width parameter and $z_0 = z(t_0)$ the z-value around which the wavepacket is centered. The contribution c_{k_0} to a given incoming plane wave [156] can then be obtained as

$$c_{k_0} = (2/\pi)^{1/4}\sqrt{\Delta z}\exp(i(k-k_0)z_0 - \Delta z^2(k-k_0)^2) \tag{5.17}$$

where Δz is the width of the wavepacket connected to α_0 by $\mathrm{Im}\alpha_0 = 1/4\Delta z^2$. The initial value of $\mathrm{Re}\alpha_0$ can be set to zero, unless one desires that the wavepacket contracts or expands (see below). We also notice that the expression for c_{k_0} is independent of $\mathrm{Re}\alpha_0$.

In actual calculations the time-dependent wave function is transformed to the "energy resolved" time-independent by the integral

$$\psi_E(x, y, z) = \frac{1}{\sqrt{2\pi}}\int dt \exp(iEt/\hbar)\psi(x, y, z, t) \tag{5.18}$$

This resolution can be carried out at a number of energies in an interval around the energy where the Gaussian wavepacket is centered, i.e., around $E_0 = \hbar^2 k_0^2/2m$. When the flux ratio is calculated at a given energy using $\psi_E(x, y, z)$ and eq. (5.10), it is simply divided by the amount $|c_{k_0}|^2$ of that energy component present in the initial wavepacket [156].

For diatom surface scattering, the number of quantum numbers increases from three to six (excluding phonons) and, by using the expansion as given in eq. (5.13), we can transform the problem to a set of coupled equations for 2D wavepackets (in r and z) labeled by four quantum numbers (n, m, j, m_j). However, with present computer facilities, the number of 2D wavepackets which can be propagated simultaneously is limited to about 1–200. Hence, in order to, for instance, include phonon and electron hole pair coupling, the scheme must be simplified even further. This is conveniently done by introducing classical mechanics for some degrees of freedom. Further reduction in the complexity can also be introduced by assuming a time-dependent self-consistent field (TDSCF) approach where the wavefunction is assumed to be of the type [168]

$$\psi(X, Y, Z, r, \theta, \phi, \{Q_k\}, t) \sim \psi_1(Z, r, t)\psi_2(X, Y, \theta, \phi, t)\psi_3(\{Q_k\}, t) \tag{5.19}$$

where $\{Q_k\}$ is a set of phonon coordinates. The center of mass coordinates for the diatomic molecule have been introduced as X,Y, and Z. In the TDSCF treatment this "trial" function is inserted in the TDSE and an equation in one set of coordinates is obtained by averaging over the others. Thus the degrees of freedom in that particular set only feels an average of the dynamics in the others. Such a treatment is expected to be adequate when the coupling between the sets of coordinates is small. Thus, strongly coupled degrees of freedom should therefore be bunched together.

A method which has been used in several studies [155,157] involves a time-dependent wavepacket propagation in the two-dimensional space Z,r using a grid representation of the wavepacket [156], a classical mechanical treatment of the variables X,Y,θ, ϕ together with the forced quantum oscillator treatment of the phonon problem (see Chapter 8).

The wavefunction $\psi_1(Z, r, t)$ is initialized as

$$\psi_1(Z, r, t_0) = \phi_n(r)\Phi_{GWP}(Z, t_0) \tag{5.20}$$

where $\phi_n(r)$ is a vibrational eigenfunction for the diatomic molecule and

$$\Phi_{GWP}(Z, t_0) = N \exp\left(-\frac{(Z - Z_{foc} - \hbar k_0 t_0/M)^2}{4(\Delta Z)^2 - 2i\hbar t_0/M} - ik_0 Z\right) \tag{5.21}$$

where M is the mass of the diatom, $\hbar k_0$ the average initial center of mass velocity, $\Delta Z = Z - Z_0$ the width of the wavepacket, N a normalization factor

$$N = 1/\sqrt{\Delta Z\sqrt{2\pi}} \tag{5.22}$$

Z_{foc} is the focusing distance, and $t_0 = (Z_0 - Z_{foc})/\hbar k_0$ the focusing time. This wavepacket contracts before it expands. This can, in some cases, be convenient in order to prevent the packet reaching the grid edge before being scattered from the surface. At the boundaries of the grid, the dissociative flux $F_d(t, r^*)$ can be obtained as

$$F_d(r^*, t) = \int dZ \frac{i\hbar}{2m}(\psi_1(Z, r, t)\frac{\partial\psi_1(Z, r, t)^*}{\partial r} - \psi_1(Z, r, t)^* \frac{\partial\psi_1(Z, r, t)}{\partial r}) |_{r=r^*} \tag{5.23}$$

The flux is calculated at a point r^* just before the wavepacket is absorbed by an imaginary potential.

Likewise at the Z edge of the grid (see Fig. 5.1), we can obtain the flux $F_r(t, Z^*)$ for the reflected wave as

$$F_r(Z^*, t) = \int dr \frac{i\hbar}{2M}(\psi_1(Z, r, t)\frac{\partial\psi_1(Z, r, t)^*}{\partial Z} - \psi_1(Z, r, t)^* \frac{\partial\psi_1(Z, r, t)}{\partial Z}) |_{Z=Z^*} \tag{5.24}$$

where m is the reduced and M the mass of the diatomic molecule.

The dissociation and reflection probabilities are obtained as time integrals over the above expressions for the flux, i.e.,

$$P_{diss} = \int dt F_d(r^*, t) \tag{5.25}$$

$$P_{refl} = \int dt F_r(Z^*, t) \tag{5.26}$$

For $t \to \infty$ we have $P_{diss} + P_{refl} = 1$. If phonon coupling is included, we can, however, have part of the wavefunction survive at the surface, and hence

$$P_{ads} = \lim_{t\to\infty} \int_0^{r^*} dr \int_0^{Z^*} dZ |\psi_1(Z, r, t)|^2 \tag{5.27}$$

has to be added to obtain unity, i.e., $P_{diss} + P_{refl} + P_{ads} = 1$.

The semiclassical wavepacket dynamics, where a few degrees of freedom are quantized and the remaining treated "classically" [74], is convenient to apply if a dynamical treatment including the full coupling to surface corrugation and excitation processes is sought. Approximate ways of including these aspects can be obtained through the Transition State Theory or Reaction Path Theory. These methods can be applied if the potential surface exhibits a transition state, or if a reaction path can be found (see [156] and Appendix A).

5.1 DISSIPATIVE QUANTUM METHODS

The dynamical methods used for treating atom-surface dynamics should be able to include the effect of changing the surface temperature. Thus the average energy transfer (the energy accommodation) should depend upon the surface temperature. Experimentally one finds an energy accommodation

$$\Delta E \sim \beta(T - T_s) \tag{5.28}$$

where T_s is the surface and T the gas temperature. That is, if the surface temperature is large, then the energy transfer is negative.

The detailed-balance corrected quantum-classical method described in section 8.2 is capable of giving this surface temperature dependence [62]. More rigorous methods involve the solution of the Liouville–von Neumann equation [159]. However, this appears to be close to impossible for systems involving the number of atoms and processes of interest for molecule-surface scattering [160]. At least this is so unless approximations are introduced, for instance, combining classical and quantum mechanical treatments or self-consistent field approximations [161].

An interesting alternative is the semigroup theory of Lindblad, Davies, and Kossakowski [162,165,163]. This and similar theories have been used in molecule surface scattering by Kosloff [164] and Jackson [166].

In order to incorporate the surface temperature dependence we need to introduce an ensemble average over the states of the solid. Thus, introducing the state vector $|\Psi_{\overline{n}}(t)\rangle$, we obtain the density operator for the total system

$$\rho = \frac{1}{Z} \sum_{\overline{n}} \exp(-E_{\overline{n}}/kT_s) |\Psi_{\overline{n}}(t)\rangle\langle\Psi_{\overline{n}}(t)| \tag{5.29}$$

where

$$|\Psi_{\bar{n}}(t)\rangle = |\psi_{n_0}\psi_{n_1} \ldots \psi_{n_M}\rangle \tag{5.30}$$

Z the partition function ($Z = tr(\rho)$), $n_i (i = 1, \ldots M)$ denotes bath or reservoir (R) states, and n_0 the quantum state of the system (S). A reduced density matrix for the S system is obtained as

$$\sigma_{n_0,n'_0} = tr_R \langle\psi_{n_0}|\rho|\psi_{n'_0}\rangle \tag{5.31}$$

where the trace (tr) is taken over the reservoir only. For a molecule surface system we would initially prepare the system in a product state $\psi^S_{n_0}\Psi^R_{n_1 \ldots n_M}(t_0)$. As the two systems, the molecule (S) and the surface (R), start to interact, the product type wavefunction is no longer the exact solution. By introducing the trace over the bath states we have, in principle, reduced the problem to one involving the system S. However, in order to do so we need to know the state function for the complete system at all times t and therefore no reduction in the problem has been achieved as yet. A reduction in the complexity would certainly be obtained if we could assume a product (Hartree-type) wavefunction also at later times t. However, even simpler schemes have been suggested. The effect of the bath is to add a dissipative term to the equations of motion. In the semigroup theory mentioned above, the form of the dissipative terms is obtained by introducing the concept of "completely positive mapping." The theory is fairly formal and will not be discussed in great detail here. For more details, the reader is referred to the literature cited above. In the next chapters we shall instead focus on simpler theories, which also define a dissipative force.

The semigroup theory results in a quantum mechanical Markovian master equation for the evolution of the density operator ρ. Cerjan and Kosloff [164] introduced the dissipative Liouville operator [165], which in the Heisenberg representation yields the equations of motion for an operator O as

$$\frac{d}{dt}O = \frac{i}{\hbar}[H, O] + \frac{\partial O}{\partial t} + L_D(O) \tag{5.32}$$

where the last term is the dissipative correction to the Heisenberg equation of motion. Introducing an interaction term between the two systems, the molecule and the bath (the solid), as

$$H_I = \sum_n \Gamma_n a_n b_n \tag{5.33}$$

where Γ_n is the interaction strength and a_n an operator for the system and b_n for the bath, the form of the dissipative term in the semigroup theory is given as [162]

$$L_D(O) = \sum_n \gamma_n \left(a_n O a_n^+ - \frac{1}{2}\{a_n a_n^+, O\} \right) \tag{5.34}$$

where $\{A, B\} = AB + BA$ denote the anticommutator and the γ_n's are parameters entering the interaction between the system and the bath. We notice that only the system and interaction operators are incorporated in the equations of motions. The semigroup theory is based upon formal grounds using CP (completely positive) mapping and is therefore general for a large class of quantum Markovian processes. However, the practical problem which arises is, what the values of the parameters γ_n are, and also which system operators a_n one should incorporate in the dissipative term.

On less formal grounds a Markovian type of equation was also obtained by Redfield [167]. Redfield used perturbation theory up to second order and derived a Master equation for the evolution of the density matrix $\sigma_{\alpha\alpha'}$, where α, α' denote quantum states of the molecule with energy defined by $\hbar\alpha$. The simple equation for the relaxation of the density matrix was obtained by neglecting the ensemble-averaged first-order terms. Thus we have [167]

$$\frac{d}{dt}\sigma_{\alpha\alpha'} = i(\alpha - \alpha')\sigma_{\alpha\alpha'} + \sum_{\beta\beta'} R_{\alpha\alpha'\beta\beta'}\sigma_{\beta\beta'} \tag{5.35}$$

where $\mathbf{R}$ is a relaxation matrix, the elements of which are related to a correlation function

$$C_{\alpha\alpha'\beta\beta'}(\tau) = \langle G_{\alpha\beta}(t)G_{\alpha'\beta'}(t - \tau)\rangle \tag{5.36}$$

where the brackets $\langle\rangle$ indicate an ensemble average and $G(t)$ is a random Hermitian perturbation of the molecule. The perturbation depends upon the interaction between the heat bath and the molecule and the ensemble average is carried out over a distribution of solid states. The connection between the relaxation matrix and the correlation functions is [167]

$$\begin{aligned} R_{\alpha\alpha'\beta\beta'} = {} & J_{\alpha\alpha'\beta\beta'}(\beta' - \alpha') + J_{\alpha\beta\alpha'\beta'}(\alpha - \beta) - \delta_{\alpha'\beta'}\sum_{\gamma} J_{\gamma\beta\gamma\alpha}(\gamma - \alpha) \\ & - \delta_{\alpha\beta}\sum_{\gamma} J_{\gamma\alpha'\gamma\beta'}(\alpha' - \gamma) \end{aligned} \tag{5.37}$$

where

$$J_{\alpha\alpha'\beta\beta'}(\omega) = \int_0^{\infty} d\tau \exp(i\omega\tau)C_{\alpha\beta\alpha'\beta'}(\tau) \tag{5.38}$$

This Markovian type of master equation for the time-evolution of the reduced-density matrix has been used by Jackson [106] to study energy accommodation for atom scattering of a Cu-surface. Introducing the interaction $G(t)$ as (see also Chapter 8)

$$G(t) = \sum_{k} b_k V_k^{(1)}(a_k \exp(-i\omega_k t) + a_k^{+} \exp(i\omega_k t)) \tag{5.39}$$

where a_k^+ and a_k are the phonon creation/annihilation operators, ω_k the phonon frequency, $V_k^{(1)}$ the first derivative of the interaction potential with respect to the phonon coordinate in mode k, and b_k a coupling parameter, with this expression we get

$$\langle G(t)G(t-\tau)\rangle = \Big\langle \sum_{kk'} b_k b_{k'} V_k^{(1)}(t) V_{k'}^{(1)}(t-\tau)(a_k \exp(-i\omega_k t) + a_k^+ \exp(i\omega_k t)) \times (a_{k'} \exp(-i\omega_{k'}(t-\tau)) + a_{k'}^+ \exp(i\omega_{k'}(t-\tau)))\Big\rangle \tag{5.40}$$

Introducing the ensemble average we finally obtain

$$\langle G(t)G(t-\tau)\rangle = \sum_k \frac{\hbar}{2\omega_k} V_k^{(1)}(t) V_k^{(1)}(t-\tau)(\langle n_k\rangle \exp(i\omega_k\tau) + (\langle n_k\rangle + 1)\exp(-i\omega_k\tau)) \tag{5.41}$$

where we have used the fact that the nonzero ensemble-average values are $\langle a_k^+ a_{k'}\rangle = \delta_{kk'}\langle n_k\rangle$ and $\langle a_k a_{k'}^+\rangle = \delta_{kk'}(\langle n_k\rangle + 1)$. The average value of the phonon quantum number is $\langle n_k\rangle = z_k/(1 - z_k)$, where $z_k = \exp(-\hbar\omega_k/k_B T_s)$ (see also Chapter 8). We notice that in this formulation, the motion of incoming molecule is coupled to an average over the surface modes (a heat bath) rather than to individual modes.

The Redfield theory is based on a weak coupling assumption and has been used extensively in nuclear magnetic resonance (NMR) theory for treating the influence of the surroundings on the spectra. Attempts to go beyond the weak coupling limit assumed in the Redfield theory have been made [78], and the requirement of complete positiveness assumed by Lindblad in the coupling between the subsystem and the reservoir has been challenged by Pechukas [79].

The dynamics of the system is irreversible in time, both in the Redfield and the Lindblad formalism. This irreversibility arises from the loss of phase information when averaging over the bath system.

5.2 SIMPLE MODELS FOR ATOM-SURFACE SCATTERING

Beam experiment with angular resolved time-of-flight spectra of monoenergetic He beam have provided insight in the phonon dynamics and excitation as a function of incident energy and surface temperature [49]. The experimental data follow closely an expression derived using different theoretical approaches by Gumhalter and Bilic [119], Brako and Newns [39], and Billing [40]. Thus it was shown that the intensity of the scattered He atoms in a certain energy range and angular range could be expressed as a product of three Gaussian distributions

$$I \sim \frac{1}{(2\pi)^{3/2}\sigma_x\sigma_y\sigma_0} \exp[-(\varepsilon - E_{int})^2/(2\sigma_0^2) - \Delta K_x^2/(2\sigma_x^2) - \Delta K_y^2/(2\sigma_y^2)] \tag{5.42}$$

where ε is the energy of the scattered atom, E_{int} is a quantity to be defined in Chapter 8 as $\sum_k \hbar\omega_k \rho_k$, where ω_k is the phonon frequency and ρ_k related to the probability for exciting phonon mode k. At zero Kelvin, E_{int} is the average energy transfer to the solid. We notice that the quantity is positive. The width of the energy distribution is given by

$$\sigma_0^2 = 2k_B T_s E_{int} \tag{5.43}$$

i.e., proportional to the energy transfer. Thus we see that increasing the surface temperature T_s gives rise to a broader distribution. Likewise more energetic beams will increase E_{int} and hence give a broader distribution. ΔK_x and ΔK_y involves the scattering angles (θ_f, ϕ_f) as well as the incident ones (θ_i, ϕ_i). For the surface-projected wavevectors we then have (see also Chapter 2)

$$\Delta K_x = k_f \sin\theta_f \cos\phi_f - k_i \sin\theta_i \cos\phi_i \tag{5.44}$$

$$\Delta K_y = k_f \sin\theta_f \sin\phi_f - k_i \sin\theta_i \sin\phi_i \tag{5.45}$$

where k_i and k_f are the initial and final wavenumbers of the He-atom. The corresponding width parameters are

$$\sigma_x^2 = \frac{2k_B T_s}{\hbar} \sum_k \frac{\rho_k}{\omega_k} (Q_x^{(k)})^2 \tag{5.46}$$

$$\sigma_y^2 = \frac{2k_B T_s}{\hbar} \sum_k \frac{\rho_k}{\omega_k} (Q_y^{(k)})^2 \tag{5.47}$$

i.e., a sum over phonon modes and their contribution to the parallel surface momentum $Q_x^{(k)}$ and $Q_y^{(k)}$ (see also Chapter 2). If we express $Q_x^{(k)}$ as ω_k/v_R, where v_R is some average surface atom velocity, we get $\sigma_x^2 = 2kT_s E_{int}/(\hbar^2 v_R^2)$ [38], i.e., that also the width in the two other Gaussians are proportional to the average energy transfer E_{int}. The above simplification can be justified by noting that for surface waves have $\omega \sim \sqrt{4F/m_\alpha}(1/2)Qa$ where a is the lattice constant, Q the surface wave number, F the force constant and m_α the mass of a surface atom (see Chapter 2). Thus introducing $F = m_\alpha \omega^2$ and $v_R = a\omega$ where v_R is the propagation velocity of a wave traveling along the surface, we get $F \sim m_\alpha v_R^2/a^2$ and hence $\omega \sim Q v_R$. We notice that the expression (5.42) does not have its maximum for $\varepsilon = E_{int}$, which incidently would always be positive. Rather the position (see exercise 5.2) depends on the scattering angle due to the energy conservation $(\hbar^2 k_f^2/2m) = (\hbar^2 k_i^2/2m) + \varepsilon$, where ε is the energy exchange with the surface, negative for energy loss to the phonons and positive for energy gain (see Fig. 5.2). We notice that the expression for the intensity (5.42) through the factor $\sigma_x\sigma_y\sigma_0$ follows a $T_s^{-3/2}$ dependence at high temperatures, in agreement with experimental findings [37,80].

Various other simple semiclassical models both for atom- and molecule-surface diffraction and energy transfer have been proposed by Miller and coworkers [51]

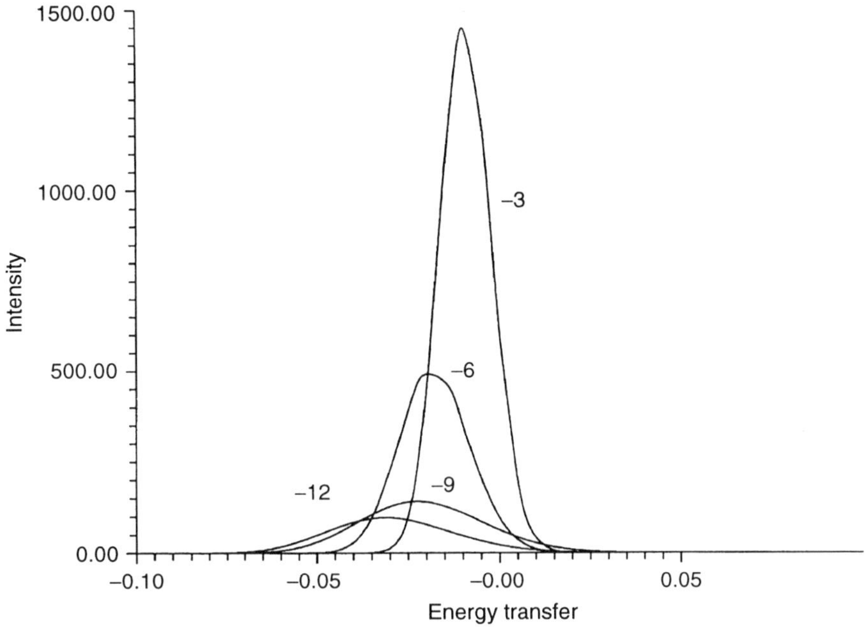

FIGURE 5.2 Calculated scattering intensities as a function of energy transfer ε (in kJ/mol) to a copper surface by He-atom collisions at various initial approach angles θ_i. Negative values of ε denote energy loss to the phonons and positive an energy gain. The initial approach angle $\theta_i = \theta_f + \Delta\theta$, where $\Delta\theta = -3, -6, -9$, and $-12°$. The initial kinetic energy is 10.9 kJ/mol and $\theta_f = 47.9°$.

and Billing [75]. If the corrugation of the surface is week, we can approximate the potential as

$$V(x, y, z) = V_0(z) + V_1(z)(\cos(2\pi x/a_x) + \cos(2\pi/a_y)) \tag{5.48}$$

Assuming a trajectory $z(t)$ for the motion toward the surface, it is possible [75,51] to derive a simple expression for the diffraction probabilities, i.e., for scattering from channel $(n_0 m_0)$ to (n, m) we have

$$P_{n_0 m_0;nm} = J_{n-n_0}(A_x)^2 J_{m-m_0}(A_y)^2 \tag{5.49}$$

where J_n is a Bessel function and the arguments Fourier transforms of the potential

$$A_x = \frac{1}{\hbar}\int_{-\infty}^{\infty} dt V_1(t)\cos(\omega_x t) \qquad (5.50).$$

with a similar expression for A_y. The frequency ω_x is defined as

$$\omega_x = 2\pi\hbar\overline{K}_x/Ma_x \tag{5.51}$$

where M is the mass of the atom and $\overline{K}_x$ is an average wavevector

$$\overline{K}_x = K_x + \frac{1}{2}(n + n_0)\frac{2\pi}{a_x} \tag{5.52}$$

with a similar expression for the y component. The magnitude of K_x and K_y are obtained from the incident angles as $K_x = k_i \sin\theta_i \cos\phi_i$ and $K_y = k_i \sin\theta_i \sin\phi_i$ (see also section 2.4). Comparison with quantum calculations on diffraction [52] show that the specular (00) and also higher-order diffraction intensities often can be reproduced quantitatively with the simple expression above [51].

EXERCISES

5.1: Consider a forced oscillator with the hamiltonian $H = \hbar\omega\left(a^+a + (1/2)\right) + f(t)(a + a^+)$ and obtain the Heisenberg equations of motion including a dissipative term for the system operators a, a^+ and a^+a.

5.2: Derive an expression for the maximum of the intensity $I(\varepsilon)$ given by eq. (5.42). Assume that $\phi_i = \phi_f$, $E_{int} = 1$ kJ/mol, $E_{kin} = 10.9$ kJ/mol, and $\sigma_x = \sigma_y$. Calculate the maximum of the energy distribution for $\theta_f = 47.9°$ and $\theta_i = 50.9°$, 44.9°, 41.9°, and 38.9° for He scattering from a Cu(001) surface. The lattice constant is $a = 2.5$ Å, and the dominating phonon frequency being excited has $\omega = 50$ cm^{-1}. Compare the energy shift, i.e., $\delta\varepsilon = \varepsilon_{max}(\theta_i) - \varepsilon_{max}(\theta_i = \theta_f)$, for the 4 values of θ_i with the experimental data, which are 8 meV, −12 meV, −21 meV, and −32 meV, respectively.

6

CLASSICAL MECHANICAL TREATMENT

In the so-called "molecular dynamics" (MD) studies one solves the classical equations of motion [18] often for a large number of atoms interacting with each other. The calculations are usually carried out assuming simple additive pair potentials. The classical mechanical treatment of the nuclear mechanics has become popular for several reasons. It is possible to solve the equations of motion for a large number of atoms (several thousands), and by introducing the so-called Monte-Carlo averaging technique it is possible with a modest number of trajectories to obtain good information on averaged quantities such as dissociation and recombination rates, energy accommodation and even internal state distribution of products. The methodology has however some limitations and problems mentioned at the end of this chapter.

The MD approach has also been extensively used in gas-surface calculations and the first calculations already appeared in the sixties [71]. Thus the atoms are divided in two groups

1. $i = 1, \ldots n$ gas-phase atoms
2. $k = 1, \ldots N$ lattice atoms

The interaction potential is, as mentioned, often assumed to be a pairwize additive potential containing terms such as $V_{11}(R_{ij})$, $V_{12}(R_{ik})$ and $V_{22}(R_{kl})$, where the indices refer to the group given above. In the harmonic approximation the latter interaction will be

$$V_{22}(R_{kl}) = \tfrac{1}{2} F_{kl} (\mathbf{R}_k - \mathbf{R}_l)^2 \tag{6.1}$$

where $\mathbf{R}_k$ is the position vector of the kth atom in the solid and F_{kl} a force constant. The total hamiltonian is then

$$H = \sum_{i=1}^{n} \sum_{\gamma=x,y,z} \frac{1}{M_i} P_{i\gamma}^2 + \sum_{k=1}^{N} \sum_{\gamma=x,y,z} \frac{1}{M_k} P_{k\gamma}^2 + \\ + \sum_{i<j} V_{11}(R_{ij}) + \sum_{k<l} V_{22}(R_{kl}) + \sum_{ik} V_{12}(R_{ik}) \tag{6.2}$$

where the first two terms are the kinetic energy of the gas-phase and surface atoms and the last the various potential energies. When calculating the potential and from that the forces on the N lattice atoms one should incorporate forces from outside the finite lattice by enforcing "periodic boundary" conditions—utilizing the information on the lattice structure.

The hamiltonian equations of motion are

$$\dot{\gamma}_i = \frac{P_{i\gamma}}{M_i} \tag{6.3}$$

$$\dot{\gamma}_k = \frac{P_{k\gamma}}{M_k} \tag{6.4}$$

$$\dot{P}_{i\gamma} = -\frac{\partial H}{\partial \gamma_i} = -\sum_{j \neq i} \frac{\partial V_{11}}{\partial R_{ij}} \frac{\partial R_{ij}}{\partial \gamma_i} - \sum_{k} \frac{\partial V_{12}}{\partial R_{ik}} \frac{\partial R_{ik}}{\partial \gamma_i} \tag{6.5}$$

and

$$\dot{P}_{k\gamma} = -\frac{\partial H}{\partial \gamma_k} = -\sum_{l \neq k} \frac{\partial V_{22}}{\partial R_{kl}} \frac{\partial R_{kl}}{\partial \gamma_k} - \sum_{i} \frac{\partial V_{12}}{\partial R_{ik}} \frac{\partial R_{ik}}{\partial \gamma_k} \tag{6.6}$$

where we note that

$$R_{ij} = |\mathbf{R}_i - \mathbf{R}_j| = \sqrt{(X_i - X_j)^2 + (Y_i - Y_j)^2 + (Z_i - Z_j)^2} \tag{6.7}$$

and

$$\frac{\partial R_{ij}}{\partial \gamma_i} = -\frac{\partial R_{ij}}{\partial \gamma_j} = \frac{\gamma_i - \gamma_j}{R_{ij}} \quad (\gamma = x, y, z) \tag{6.8}$$

The coupled equations (6.3)–(6.6) can be solved using standard integrators such as Runge-Kutta, Predictor-Corrector, and so on. However, using eqs. (6.3) and (6.5) we can obtain equations of the type

$$M_i \ddot{\gamma}_i = -\frac{\partial H}{\partial \gamma_i} = F_i \tag{6.9}$$

where F_i is the force on particle i. The solution to this equation can be expressed in a numerically stable time-symmetric fashion [197] as

$$\gamma_i(t+\Delta t) = -\gamma_i(t-\Delta t) + 2\gamma_i(t) + \frac{\Delta t^2}{M_i} F_i(t) \tag{6.10}$$

Higher-order expansions involving Δt^4, Δt^6, and so on may easily be derived.

The potential for the solid, i.e., V_{22}, is usually just taken in its harmonic approximation including, say, nearest and next-nearest neighbor interaction. The potential may then be expanded to second order in the displacement coordinates as

$$V = \frac{1}{2}\sum_{kl} V_{22}(R_{kl}) = V_0 + \frac{1}{2}\sum_{k\gamma}\sum_{l\delta} A_{k\gamma;l\delta}\eta_{k\gamma}\eta_{l\delta}/\sqrt{M_k M_l} \tag{6.11}$$

where $\eta_{k\gamma} = \sqrt{M_k}(\gamma_k - \gamma_k^0)$ are mass-weighted coordinates. The hamiltonian for the solid is then in the displacement coordinates

$$H_c = V + \frac{1}{2}\sum_{k\gamma} \dot{\eta}_{k\gamma}^2 \tag{6.12}$$

where γ_k^0 is the equilibrium value and $\gamma = x$, y, or z. Through an orthorgonal transformation with a matrix $\mathbf{T}$ we may diagonalize (6.12) i.e., we introduce the transformation to normal mode coordinates Q_r

$$\eta_{k\gamma} = \sum_r T_{k\gamma;r} Q_r \tag{6.13}$$

where the summation runs over the elements of the eigenvectors entering the transformation matrix $\mathbf{T}$. We then obtain

$$H_c = \sum_{r=1}^{3N-6} H_r + V_0 \tag{6.14}$$

where

$$H_r = \frac{1}{2}(\dot{Q}_r^2 + \omega_r^2 Q_r^2) \tag{6.15}$$

and $Q_r (r = 1, \ldots 3N-6)$ are the normal mode coordinates. The six remaining frequencies are $\omega_r = 0$ $(r = 3N-5, \ldots 3N)$, corresponding to translational and rotational motion of the crystal as a whole. The solution of the equations of motion for an isolated oscillator with energy E_r is

$$Q_r = \frac{\sqrt{2E_r}}{\omega_r}\cos(\omega_r t + \phi_r^0) \tag{6.16}$$

where ϕ_r^0 is an arbitrary phase angle $\phi_r^0 \in [0 : 2\pi]$. The distribution of the vibrational quanta follows the Bose-Einstein (BE) statistics, i.e., the average occupation number of mode r is

$$\langle n_r \rangle = \frac{1}{\exp(\beta\varepsilon_r) - 1} \tag{6.17}$$

and the average energy

$$\langle E \rangle = E_0 + \sum_{r=1} \hbar\omega_r \langle n_r \rangle \tag{6.18}$$

where $\beta = 1/k_B T_s$, T_s being the surface temperature and E_0 the zero-point energy

$$E_0 = \frac{1}{2}\sum_r \hbar\omega_r \tag{6.19}$$

The classical equations of motion are solved subject to initial conditions. For the solid, we may choose the phase angle ϕ_r^0 randomly and also the energy of a given oscillator may be chosen—but according to the BE distribution $f_r = (\exp(\beta\varepsilon_r) - 1)^{-1}$, which takes its maximum value for $\varepsilon_r = \varepsilon_r^0 = (1/2)\hbar\omega_r$. Thus we may define

$$\xi_r = (\exp(\beta\varepsilon_r^0) - 1)/(\exp(\beta\varepsilon_r) - 1) \tag{6.20}$$

and choose ξ_r randomly between 0 and 1. Eq. (6.20) then gives

$$\varepsilon_r = k_B T_s \ln(1 + (\exp(\beta\varepsilon_r^0) - 1)/\xi_r) \tag{6.21}$$

However, since anharmonic terms will redistribute the energy among the oscillators, one may use the simpler equipartition theorem [98], i.e., the theorem stating that the average energy in a degree of freedom in phase space is $k_B T_s/2$. Thus we have

$$\tfrac{1}{2}\langle \dot{\eta}_{k\gamma}^2 + A_{k\gamma;k'\gamma'}\eta_{k\gamma}^2/M_k \rangle = k_B T_s \tag{6.22}$$

or

$$\tfrac{1}{2}\langle P_{k\gamma}^2/M_k + F_k(R_{k\gamma} - R_{k\gamma}^0)^2 \rangle = k_B T_s \tag{6.23}$$

With the latter expression we avoid calculating the transformation matrix $\mathbf{T}$ and simply get for $t = 0$

$$R_{k\gamma} = R_{k\gamma}^0 + \sqrt{2k_B T_s/F_k}\cos(\phi_{k\gamma}^0) \tag{6.24}$$

and

$$P_{k\gamma} = \sqrt{2M_k k_B T_s}\sin(\phi_{k\gamma}^0) \tag{6.25}$$

where $R^0_{k\gamma}$ is the equilibrium position, F_k the force constant for atom k, and ϕ^0_k a random phase angle in the range 0 to 2π.

For an incident gas-phase atom with kinetic energy E_{kin}, we have the initial condition given as

$$X = R \sin\theta \cos\phi + \Delta X \quad (6.26)$$

$$Y = R \sin\theta \sin\phi + \Delta Y \quad (6.27)$$

$$Z = R \cos\theta \quad (6.28)$$

where θ is the angle between R and the z-axis, ϕ the angle between the x-axis and the projection of R in the $X - Y$ plane. ΔX and ΔY specify the aiming point (see Fig. 6.1), i.e., the point at the surface which would be hit if there was no interaction. Likewise we have for the momenta

$$P_X = -P_R \sin\theta \cos\phi \quad (6.29)$$

$$P_Y = -P_R \sin\theta \sin\phi \quad (6.30)$$

$$P_Z = -P_R \cos\theta \quad (6.31)$$

where $E_{kin} = P_R^2/2M$.

The number of coupled equations are $6(N + n)$ and the quantities calculated are obtained as averages over a number of trajectories which initially have randomly chosen values for the phase angles $\phi^0_{k\gamma}$ and the aiming points ΔX, ΔY. The trajectories are run on the lower adiabatic potential surface for the interaction between the atom/molecule and the solid. It is possible to include in a simple but approximate manner the effect of electronic nonadiabatic effects if the potential energy surfaces for the excited states are available. This is done by

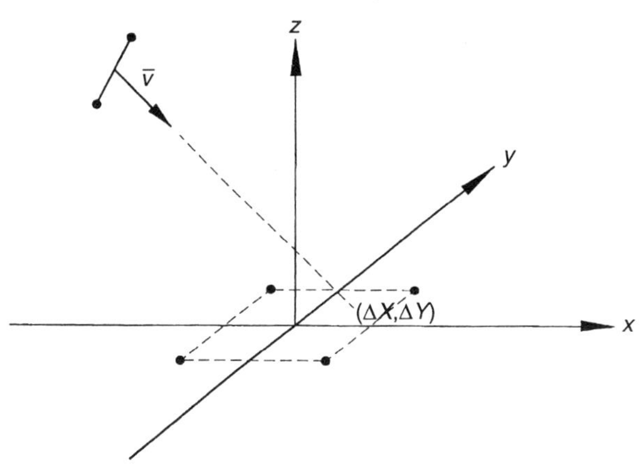

FIGURE 6.1 The aiming area on the surface. The incoming molecule hits the surface randomly in this area.

computing (along the trajectories) the probability for jumping to another surface. This probability can be estimated in various ways [72] and is large near the avoided crossings between the adiabatic surfaces. At the avoided crossing, the trajectory is allowed to jump to the upper surface according to the transition probability at that time. After the instantaneous jump, the velocities are adjusted [73] so as to conserve energy, and the integration is continued on the upper surface until the next avoided crossing is reached, and so on.

Quantities which are available from classical trajectory studies of atom-surface scattering are scattering angles, energy accommodation, surface effects (corrugation, temperature, etc.), sticking probabilities, mobility, diffusion, and so on. The approach is straightforward to use, but rather heavy computationally, due to the large number of atoms needed for a realistic modeling of the solid. In an attempt to reduce the problem to a more manageable one it has been suggested to use an approach based on the generalized Langevin equation (GLE) (see Chapter 7).

For a diatomic molecule, we need also to initialize the internal degrees of freedom. Thus the center of mass motion is initialized as for the atom, i.e., $\mathbf{R}$ is the vector from the aiming point at the surface to the center of mass of the molecule and $\mathbf{P}_R$ the momentum vector in that same direction. The vector joining the two atoms $\mathbf{r}$ is given by

$$\mathbf{r} = r(\sin\theta' \cos\phi', \sin\theta' \sin\phi', \cos\theta') \tag{6.32}$$

where r is the bond distance and θ', ϕ' specify the orientation in space. The momentum of the internal motion is defined as

$$\mathbf{p} = m\dot{\mathbf{r}} \tag{6.33}$$

where m is the reduced mass $m = m_1 m_2/(m_1 + m_2)$ of the diatomic molecule. The momentum $\mathbf{p}$ is split in a component along the bond-axis and perpendicular to the axis. Thus we have

$$\mathbf{p} = \mathbf{p}_r + \mathbf{p}_\perp \tag{6.34}$$

where the orientation of the two vectors is given by [130]

$$\mathbf{p}_r = p_r(\sin\theta' \cos\phi', \sin\theta' \sin\phi', \cos\phi') \tag{6.35}$$

for the vibrational component and

$$\mathbf{p}\perp = \frac{\hbar\left(j + \frac{1}{2}\right)}{r}(\sin\phi' \cos\eta + \cos\theta' \cos\phi' \sin\eta, \\ -\cos\phi' \cos\eta + \cos\theta' \sin\phi' \sin\eta, \sin\theta' \sin\eta) \tag{6.36}$$

for the rotational component. The rotational quantum number is j and the $+(1/2)$ added is the so-called Langer correction, i.e., $j(j+1)$ is replaced by $(j +$

$(1/2))^2$ [94]. The vibrational energy is given as

$$E_{vib} = \frac{p_r^2}{2m} + V_{int}(r) \tag{6.37}$$

where V_{int} is the intramolecular potential, for example, a Morse potential. The rotational energy is

$$E_{rot} = \frac{p_\perp^2}{2m} = \frac{\hbar^2\left(j+\frac{1}{2}\right)^2}{2mr^2} \tag{6.38}$$

The angles θ', ϕ', and η are chosen randomly and the position coordinates and momentum components of the two atoms given as:

$$\mathbf{R}_1 = \mathbf{R} - \frac{m_2}{m_1 + m_2}\mathbf{r} \tag{6.39}$$

$$\mathbf{R}_2 = \mathbf{R} + \frac{m_1}{m_1 + m_2}\mathbf{r} \tag{6.40}$$

$$\mathbf{P}_1 = \frac{m_1}{m_1 + m_2}\mathbf{P} - \mathbf{p} \tag{6.41}$$

$$\mathbf{P}_2 = \frac{m_2}{m_1 + m_2}\mathbf{P} + \mathbf{p} \tag{6.42}$$

The diatomic molecule is initially given a vibrational and a rotational quantum number v and j. Furthermore, R is chosen sufficiently large for the value of the interaction between the surface and molecule to be negligible. The kinetic center of mass energy specifies P_R and the following 6 variables are chosen randomly, θ', ϕ', η, ΔX, ΔY and κ, which specifies the bond length. The angles θ' and η are chosen in the range 0 to π, ϕ' between 0 and 2π and κ between 0 and 1. The values of ΔX and ΔY are chosen within the surface of a unit cell of the crystal.

For the Morse oscillator with the potential $D_e(1 - \exp(-\beta(r - r_e)))^2$, where r_e is the equilibrium distance, we have [130]

$$r = r_e + \frac{1}{\beta}\ln\left(-\frac{b}{2a}(1 - \sqrt{1 - 4ac/b^2}\cos(\pi\kappa)\right) \tag{6.43}$$

where $a = E_{vj} - D_e - Aj_r^2$, $b = 2D_e - Bj_r^2$, $c = Cj_r^2 - D_e$, $j_r = \hbar\sqrt{j(j+1)}$, $E_{vj} = E_{vib} + E_{rot}$, and the constants A, B, C given by

$$A = \frac{1}{2mr_e^2}\left(1 - \frac{3}{\beta r_e}\left(1 - \frac{1}{\beta r_e}\right)\right) \tag{6.44}$$

$$B = \frac{2}{m\beta r_e^2}\left(1 - \frac{3}{2\beta r_e}\right) \tag{6.45}$$

$$C = \frac{1}{2\beta mr_e^2}\left(1 - \frac{3}{\beta r_e}\right) \tag{6.46}$$

The analyses of the trajectories is carried out on the basis of the possible events. For atom-surface scattering we have two possibilities:

1. Reflection (the atom leaves the surface)
2. Sticking (the atom sticks to the surface)

The criterion for sticking is not unique—it depends on the time scales considered. The trajectories are usually followed on a time scale of the order of picoseconds, at most, say, 100 ps. On the time scale we may define sticking by monitoring the energy transfer to the solid (E_{int}). Once the energy is transferred to the surface atoms it is efficiently dissipated to the bulk of the solid; i.e., a criterion for sticking could be that $E_{int} > E_{kin} + V$ where E_{kin} is the initial kinetic energy and V the potential energy where the atom is trapped $V < 0$. For diatomic molecules we also have the possibility of:

3. Dissociation

i.e., the bond distance r exceeds some critical value r^* taken to be the smallest distance between two sites at the surface for atomic sticking. The dissociation event may be split in that:

1. Both atoms are adsorbed.
2. One atom is adsorbed, the other leaves the surface.
3. Both atoms leave the surface.

Also the reflected trajectories may be analyzed further. Thus the scattering angles (θ_f, ϕ_f) are of interest both for atoms and molecules. For molecules we furthermore may be interested in the final rotational and vibrational state. The rotational angular momentum is obtained as $\mathbf{j} = \mathbf{r} \times \mathbf{p}$ and the vibrational action n (the classical analog of the quantum number) can be obtained using the Bohr-Sommerfield quantization method

$$\left(n + \frac{1}{2}\right) = \frac{1}{h} \oint dr p_r \tag{6.47}$$

where p_r is defined through

$$\frac{p_r^2}{2m} = E - V_{int}(r) - E_{rot} \tag{6.48}$$

where E is the total energy of the molecule.

For polyatomic molecules the initialization and analyses of the classical variables is more problematic, the reason being that the hamiltonian is usually not separable, and hence it is much more complicated to determine how the energy is distributed in the various degrees of freedom or how to "quantize" the trajectories. However, a number of approximate quantization schemes do exist [95].

One interesting and potentially promising method is to divide the hamiltonian in two parts

$$H = H_0 + V \tag{6.49}$$

where H_0 is a separable hamiltonian for the system. Thus we can define the quantum numbers using, for instance, the Bohr-Sommerfeld quantization scheme

$$\left(n_k + \frac{1}{2}\right) = \frac{1}{h} \oint p_k(E, Q_k) dQ_k \tag{6.50}$$

where the integral is carried out over a closed path in phase space. Thus according to this scheme, the area in $p - Q$ space is quantized. If we assume that H_0 can be written as

$$H_0 = \sum_k \left(\frac{p_k^2}{2} + v(Q_k) \right) \tag{6.51}$$

we can use this scheme for each vibrational degree of freedom. The potential term V contains, however, coupling elements which couple the various modes. Rotational/vibrational coupling terms may be dealt with in a similar manner and included in V. In order to incorporate the term V we may use Ehrenfest's adiabatic theorem [96]; i.e., the potential V is switched on infinitely slowly by introducing a time-dependent function $f(t)$, where $f(0) = 0$ and $f(\infty) = 1$. The theorem tells you that the quantum numbers are unchanged by this procedure but the energy of the state is changed. This scheme has been used for quantizing nonseperable systems in a number of recent publications [95].

The classical mechanical molecular dynamics approach is usually straightforward to use but has a number of problems, some of which are often overlooked. Obviously, inherent quantum effects are not included in the dynamics—here we mention classically forbidden events, such as tunneling or weak transitions between various "quantum states," diffraction, nonadiabatic electronic effects which must be incorporated and added to the theory in an approximate way. There are problems in assigning quantum numbers to the various trajectories, especially for polyatomic molecules, and even if some procedure can be found, the state-resolved quantities may not be very accurate. A big problem is the conservation of vibrational zero-point energy, not only in the solid, but also for the gas-phase molecules. This problem increases if the interaction between the molecule and the surface is strong [97]. However, despite these serious shortcomings the MD approach can often, through straightforward calculations, give a good estimate of reaction rates, mechanisms, and energy accommodation, and so on. The use of Monte-Carlo sampling techniques makes it convenient to model efficiently experimental conditions such as those obtained in nozzle beam experiments for which one may introduce a statistical sampling over the Gaussian distribution function of the initial center of mass velocity.

EXERCISES

6.1: Calculate using eq. (6.43) the turning points for the vibrational motion of hydrogen in $v = j = 0$ using the facts that $\beta = 1.878$ Å^{-1} and

$$E_{vj} = \hbar\omega_e\left(v+\frac{1}{2}\right) - \hbar\omega x_e\left(v+\frac{1}{2}\right)^2 + j(j+1) \times \left(B_e - D_v j(j+1) - \alpha_e\left(v+\frac{1}{2}\right)\right) \quad (6.52)$$

where $B_e = 1/2mr_e^2$, $D_v = \hbar^4/(4m^2\beta^2 r_e^6 D_e)$ and $\alpha_e = (3/4)\hbar^3\omega_e(1 - 1/\beta r_e)/(m\beta r_e^3 D_e)$. m is the reduced mass, and for hydrogen we have D_e=4.7477 eV, $\omega_e = 4395$ cm^{-1}, $x_e = 0.02685$, and $r_e = 0.7417$ Å.

6.2: Use the Bohr-Sommerfeld quantization to derive an expression for the vibrational/rotational energies E_{vj} of a harmonic oscillator. Compare with the above expression for the Morse oscillator. Use the fact that

$$\int dx\sqrt{a+bx+cx^2} = \frac{2cx+b\sqrt{a+bx+cx^2}}{4c} + \frac{-\Delta}{8c\sqrt{-c}}\arcsin((2cx+b)/\sqrt{-\Delta}) \quad (6.53)$$

where $c < 0$ and $\Delta = 4ac - b^2 < 0$.

7

THE GENERALIZED LANGEVIN EQUATION

The solution of the many-body problem as defined in the previous section on classical mechanics, is numerically demanding. Considering, furthermore, that the primary effect of the surface atoms is assumed to be that of a heat bath maintained at a certain temperature, it is naturally to think that some approximate description would suffice.

Thus the solid acts as a sink or a source to or from which energy can be transferred or received. If the effect of the surface is of this simple nature, it is conceivable that a simplified description of the motion of the surface atoms would be sufficient. In order to reduce the numerical work involved in the purely classical treatment of the surface dynamics and to introduce a simplified description, the generalized Langevin equation (GLE) has been used and formulated by Adelman [83], Tully [84], and DePristo [85]. In Chapter 2, we have considered the problem of energy transfer to a linear chain as formulated by Zwanzig [21] and seen that dissipative forces are influencing the dynamics of the incoming particle. It is conceivable that the GLE approach is accurate enough for describing the coupling to the phonons if they are excited as bulk phonon modes. If, however, information on the specific phonon mode excitation is needed, as, for instance, on how much the surface waves contribute to the overall process, then an alternative methods such as the one described in the next chapter should be considered.

In order to formulate a theory and derive the equations more generally, we divide the N atoms of the solid in two groups consisting of P primary atoms (see Fig. 7.1) and N − P secondary ones. From the hamiltonian (6.12) we get the equations of motion

$$\ddot{\eta} = -\mathbf{B}\eta \tag{7.1}$$

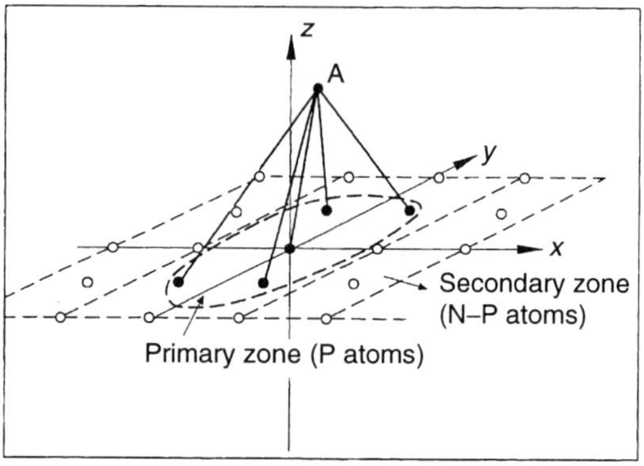

FIGURE 7.1 The atom A interacts with the primary atoms (P). A generalized Langevin equation can be derived for the dynamics of the P surface atoms.

where

$$[\mathbf{B}]_{k\gamma;l\rho} = \frac{1}{\sqrt{M_k M_l}} V_{k\gamma;l\rho} \tag{7.2}$$

η is a 3N dimensional vector, $\mathbf{B}$ a $3N \times 3N$ matrix, and components of the force constant matrix are $V_{k\gamma;l\rho} = (\partial^2 V/\partial R_{k\gamma} R_{l\rho})$. We may then partition eq. (7.1) as:

$$\begin{bmatrix} \ddot{\eta}_1 \\ \ddot{\eta}_2 \end{bmatrix} = - \begin{bmatrix} \mathbf{B}_{11} & \mathbf{B}_{12} \\ \mathbf{B}_{21} & \mathbf{B}_{22} \end{bmatrix} \begin{bmatrix} \eta_1 \\ \eta_2 \end{bmatrix} \tag{7.3}$$

where η_1 is a 3P dimensional vector containing the displacement coordinates for the P primary lattice atoms. Likewize η_2 is a 3N − 3P dimensional vector for the remaining atoms. Thus we get

$$\ddot{\eta}_1 = -\mathbf{B}_{11}\eta_1 - \mathbf{B}_{12}\eta_2 \tag{7.4}$$

and

$$\ddot{\eta}_2 = -\mathbf{B}_{21}\eta_1 - \mathbf{B}_{22}\eta_2 \tag{7.5}$$

Eq. (7.5) may now be solved formally such that:

$$\eta_2(t) = \eta_2^0(t) + \int_0^t dt' \mathbf{M}(t-t')\eta_1(t) \tag{7.6}$$

where

$$\eta_2^0(t) = \cos(\sqrt{\mathbf{B}_{22}}t)\eta_2(0) + \sin(\sqrt{\mathbf{B}_{22}}t)\mathbf{B}_{22}^{-1/2}\dot{\eta}_2(0) \tag{7.7}$$

and

$$\mathbf{M}(t-t') = -\sin(\sqrt{\mathbf{B}_{22}}(t-t'))\mathbf{B}_{22}^{-1/2}\mathbf{B}_{21} \tag{7.8}$$

If eq. (7.6) is substituted into eq. (7.4) we get

$$\ddot{\eta}_1 = -\mathbf{B}_{11}\eta_1 - \int_0^t dt'\dot{\mathbf{L}}(t-t')\eta_1(t') + \mathbf{F}(t) \tag{7.9}$$

where

$$\mathbf{F}(t) = -\mathbf{B}_{12}\eta_2^0(t) \tag{7.10}$$

and

$$\mathbf{L}(t) = \mathbf{B}_{12}\cos(\sqrt{\mathbf{B}_{22}}t)\mathbf{B}_{22}^{-1}\mathbf{B}_{21} \tag{7.11}$$

The proof of these expression may easily be found by substituting in eqs. (7.4), and (7.5). In eq. (7.9) $\dot{\mathbf{L}}(t)$ is a memory function picking up values of the solution $\eta_1(t')$ in the range $(0 \le t' \le t)$ whereas $\mathbf{F}(t)$ appears as a fluctuating random force. Eq. (7.9) is only useful if some approximate scheme for the memory function $\mathbf{L}(t)$ and the fluctuating force $\mathbf{F}(t)$ can be introduced. A memory function already appeared in section 2.3 when energy transfer to the linear chain was considered.

An important theorem known as the *fluctuation dissipation theorem* [86] connects the dissipative force $\mathbf{L}(t)$ to the ensemble average over the fluctuations through the equation

$$\langle \mathbf{F}(O)\mathbf{F}^{+}(t)\rangle = k_B T_s \mathbf{L}(t) \tag{7.12}$$

where $\mathbf{F}^{+}$ is the adjoint vector and T_s the surface temperature.

This equation is used to obtain reasonable models for the fluctuating force if the dissipation term $\mathbf{L}(t)$ is known. In order to obtain $\mathbf{L}(t)$, one uses experimental information on the phonon density of states $g(\omega)$ which is related to the cosine-transform of the velocity autocorrelation function

$$g(\omega) = \int_0^\infty \langle \dot{\eta}(t)\dot{\eta}(0)^T\rangle \cos(\omega t)\, dt \tag{7.13}$$

where T denotes a transposed vector. Using partial integration, we obtain from eq. (7.9)

$$\ddot{\eta}_1(t) = -\mathbf{B}_{11}\eta_1(t) + \mathbf{L}(t)\eta_1(0) - \mathbf{L}(0)\eta_1(t) + \int_0^t dt'\mathbf{L}(t-t')\dot{\eta}_1(t') + \mathbf{F}(t) \tag{7.14}$$

Multiplying from the right with $\dot{\eta}_1^{+}(0)$ and performing an ensemble average we get

$$\frac{d}{dt}\langle \dot{\eta}_1(t)\dot{\eta}_1^T(0)\rangle = -\mathbf{B}_{11}^{*}\langle \eta_1(t)\dot{\eta}_1(0)^{+}\rangle + \int_0^t dt'\mathbf{L}(t-t')\langle \dot{\eta}_1(t')\dot{\eta}_1(0)^T\rangle \tag{7.15}$$

where $\mathbf{B}_{11}^{*}$ is defined by

$$\mathbf{B}_{11}^{*} = \mathbf{B}_{11} + \mathbf{L}(0) \tag{7.16}$$

and where we have used the fact that

$$\langle \eta_1(0)\dot{\eta}_1^T(0)\rangle = 0 \tag{7.17}$$

and

$$\langle \mathbf{F}(t)\dot{\eta}_1^T(0)\rangle = 0 \tag{7.18}$$

which follows from eqs. (7.10). We now introduce the velocity autocorrelation function

$$\mathbf{P}(t) = \left\langle \dot{\eta}_1(t)\dot{\eta}_1(0)^T \right\rangle \tag{7.19}$$

and notice that the left-handside of eq. (7.15) is $\dot{\mathbf{P}}(t)$. The cos-transform of eq. (7.15) yields

$$\begin{aligned}\int_0^\infty dt\dot{\mathbf{P}}(t)\cos\omega t &= -\mathbf{B}_{11}^* \int_0^\infty dt \left\langle \eta_1(t)\dot{\eta}_1(0)^T \right\rangle \cos\omega t \\ &\quad + \int_0^\infty dt \int_0^t dt' \mathbf{L}(t-t')\mathbf{P}(t')\cos\omega t\end{aligned} \tag{7.20}$$

From eq. (7.20) and the corresponding sin-transform we get (see exercise 7.2)

$$C(\omega) = \int_0^\infty dt \cos\omega t \mathbf{P}(t) = \frac{\mathbf{\Lambda}_c \omega^2 \langle \dot{\eta}_1(0)\dot{\eta}_1^T(0)\rangle}{\mathbf{\Lambda}_c^2\omega^2 + (\omega^2 + \omega\mathbf{\Lambda}_s - \mathbf{B}_{11}^*)^2} \tag{7.21}$$

This equation gives a relation between the velocity autocorrelation function and the cos- and sin-transforms of the memory function $\mathbf{L}(t)$

$$\mathbf{\Lambda}_c = \int_0^\infty dt\mathbf{L}(t)\cos\omega t \tag{7.22}$$

$$\mathbf{\Lambda}_s = \int_0^\infty dt\mathbf{L}(t)\sin\omega t \tag{7.23}$$

For the experimentally available phonon density distribution we have $g(\omega)$ given by eq. (7.13). Comparing eqs. (7.13) and (7.21) we see that

$$\lim_{N_p \to N} C(\omega) = g(\omega) \tag{7.24}$$

i.e., if the number of primary atoms increases toward N one would expect to have the phonon distribution given by eq. (7.21). Considering then

$$\lim_{N_P \to N}\langle \dot{\eta}_1(0)\dot{\eta}_1(0)^+\rangle = \lim_{N_P \to N} \sum_{k=1}^{N_P} \dot{\eta}_{1k}(0)^2 \tag{7.25}$$

and using eq. (6.25) from the Chapter 6 on classical mechanics, we have

$$\dot{\eta}_{1k}(0) = \sqrt{2k_B T_s}\sin(\phi_k^0) \tag{7.26}$$

i.e.

$$\lim_{N_P \to N} \sum_{k=1}^{N_P} \dot{\eta}_k(0)^2 = k_B T_s \tag{7.27}$$

Thus this suggests that we may simply use

$$g(\omega) \sim \frac{\omega^2 \mathbf{\Lambda}_c k_B T_s}{\mathbf{\Lambda}_c^2 \omega^2 + (\omega^2 + \omega \mathbf{\Lambda}_s - \mathbf{B}_{11}^*)^2} \tag{7.28}$$

which gives a connection between $\mathbf{\Lambda}_c$, $\mathbf{\Lambda}_s$ and the experimentally available phonon distribution $g(\omega)$.

In order to solve for $\mathbf{L}(t)$ and $\mathbf{F}(t)$ one has to assume specific analytical expressions and then extract the relevant information from, for example, experimental determination of the phonon spectrum $g(\omega)$. A very simple choice for the memory function is

$$L(t) = a\delta(t) \tag{7.29}$$

In this case of no memory we get from eq. (7.9)

$$\ddot{\eta}_1(t) = -\mathbf{B}_{11}\eta_1(t) + a\delta(t)\eta_1(0) + a\dot{\eta}_1(t) + \mathbf{F}(t) \tag{7.30}$$

i.e. the acceleration of the primary atoms contains a fluctuating force $\mathbf{F}(t)$, a frictional force $a\dot{\eta}_1(t)$, and a linear "potential" force $-\mathbf{B}_{11}\eta_1(t)$. This equation is known as the Langevin equation. In order to include some memory, other functional forms have been suggested, for example

$$\mathbf{L}(t) = \mathbf{L}_0 \exp(-at)(\cos(\omega t) + b\sin(\omega t)) \tag{7.31}$$

Once the functional form for $\mathbf{L}(t)$ has been established one can use the fluctuation-dissipation theorem (7.12) to obtain a reasonable model for $\mathbf{F}(t)$. The GLE method is capable of discribing the friction fealt by the incoming molecule. Also the "thermal" respons from the bath to the primary atoms is included. The atoms in the primary zone are given velocities and positions obtained by sampling from a Boltzmann distribution at the surface temperature T_s.

The classical GLE method can be combined with a quantum mechanical description of the gas atom or molecule [87]. The two systems are coupled together by letting the surface atoms be forced by a mean field potential obtained by averaging over the gas atom coordinates [89]. Such models are often referred to classical path models. Other classical path models treat the gas atom dynamics by classical mechanics, whereas the phonons are quantized.

How well the GLE model works in practical calculations depends upon the parameters entering the memory kernel. In addition, the primary atoms have to be changed when the molecule moves at the surface, i.e., the P atoms are changed during the collision and according to the motion of the molecule, in such a way that the molecule is always within the primary zone. The GLE- equation has been used in its simplified one-dimensional version to study the influence of phonon-friction on activated surface diffusion [88].

In the semiclassical model introduced in the next chapter the shortcomings mentioned above are overcome by introducing a finite but large crystal consisting of enough atoms for obtaining convergence of the quantities of interest — as, for example, energy accommodation, sticking probability and so forth.

EXERCISES

7.1: Show that eq. (7.13) holds.

7.2: Derive eq. (7.21).

7.1 A MODEL HAMILTONIAN

For some model hamiltonians it is easy to obtain an explicit expression which tells us something about the importance of the "friction" felt by a small system, due to inelastic processes in the large system to which it is dynamically coupled. Consider, for example, the following simplified hamiltonian [90]

$$H = \frac{1}{2m} p_z^2 + v(z) + (z - z_{eq}) \sum_k c_k Q_k + \frac{1}{2} \sum_k (P_k^2 + \omega_k^2 Q_k^2) \tag{7.32}$$

where the first term is the kinetic energy of the incoming atom, $v(z)$ the interaction potential, and the last term the hamiltonian for the isolated crystal. The interaction between the incoming atom and the solid is linear in the normal mode coordinates Q_k and in this model linearized around the minimum of the physisorption well, located at z_{eq}. The equations of motion are now

$$m\ddot{z} = -v'(z) - \sum_k c_k Q_k \tag{7.33}$$

$$\ddot{Q}_k = -c_k(z - z_{eq}) - \omega_k^2 Q_k \tag{7.34}$$

where $\ddot{z} = (d^2/dt^2)z$, and so on. In order to solve these equations it is convenient to introduce a Laplace transform $L(z) = \int_0^\infty dt \exp(-st) z(t)$. Thus we get (for details, see exercise 7.3)

$$m(s^2L(z) - sz_0 - \dot{z}_0) = L(-v'(z)) - \sum_k c_k \left[\frac{sQ_k^0 + \dot{Q}_k^0}{s^2 + \omega_k^2} + \frac{c_k z_{eq}}{s(s^2 + \omega_k^2)} \right]$$
$$+ \sum_k \frac{c_k^2}{s^2 + \omega_k^2} L(z) \tag{7.35}$$

where $L(z)$ denotes the Laplace transform of z, $z_0 = z(t = 0)$, and $Q_k^0 = Q_k(t = 0)$. The last term can now be rewritten as

$$\sum_k \frac{c_k^2}{s^2 + \omega_k^2} = \sum_k \frac{c_k^2}{\omega_k^2} - s^2 \sum_k \frac{c_k^2}{\omega_k^2(s^2 + \omega_k^2)} \tag{7.36}$$

Following ref. [91], we replace the summation over the phonon modes k by an integral

$$\sum_k \longrightarrow \int d\omega g(\omega) \tag{7.37}$$

where $g(\omega)$ is the density of the phonon frequencies. According to the Debye model for crystals [98] (see section 2.3), we would have, for N oscillators,

$$g(\omega) = \frac{3N}{\omega_D^3}\omega^2 \tag{7.38}$$

for $\omega \leq \omega_D$ the Debye frequency, and 0 otherwise. Hence, in order to proceed, we assume the frequency dependence as in the Debye model and introduce

$$g(\omega_k) = \frac{2\eta}{\pi} m \frac{\omega_k^2}{c_k^2} \tag{7.39}$$

where then $3N/\omega_D^3$ is expressed as $2\eta m/(\pi c_k^2)$. Strictly speaking, some average value $\langle c_k^2 \rangle$ is used instead of c_k^2. What approximations are involved is not a trivial matter, since we start with equations which obey time or microscopic reversibility and end up with an equation which does not. With the above substitution, we then get for the last term in eq. (7.36), using that $\int_0^\infty dx/(x^2 + s^2) = \pi/2s$,

$$s^2 \sum_k \frac{c_k^2}{\omega_k^2(s^2 + \omega_k^2)} \sim \eta m s \tag{7.40}$$

Inserting in eq. (7.35) and performing the inverse Laplace transform finally yields

$$m\ddot{z} = -m\eta\dot{z} - w'(z) + f(t) \tag{7.41}$$

i.e., a Langevin equation in which the terms are directly obtainable from the original hamiltonian. Thus a friction term $\eta\dot{z}$ enters together with an effective (frequency-shifted) potential $w(z)$ and a fluctuating force. The potential $w(z)$ is given by

$$w(z) = v(z) + \frac{z^2}{2}\sum_k \frac{c_k^2}{\omega_k^2} \tag{7.42}$$

i.e., for a harmonic potential $v(z) = (1/2)m\omega^2(z - z_{eq})^2$ we see that the influence of the solid is to shift the frequency, change the equilibrium position, and add a frictional term to the force. In this model the friction coefficient is given as $\eta \sim ((3N\pi/2m)\langle c_k^2\rangle)/\omega_D^3$, i.e., proportional to the strength of the linear coupling.

The surface motion also gives rise to a fluctuating force

$$f(t) = -\sum_k c_k \left\{ Q_k^0 \cos(\omega_k t) + \frac{\dot{Q}_k^0}{\omega_k}\sin(\omega_k t)\right\} - z_{eq}\sum_k \frac{c_k^2}{\omega_k^2}(1 - \cos(\omega_k t)) \tag{7.43}$$

We see that the fluctuating force depends only on the solid motion. For tunneling processes, such as those important for diffusion or dissociation at surfaces or other activated processes, the presence of a fluctuating term can be very important for the tunneling rate. Thus the activation barriers will be fluctuating and the diffusion over a fluctuating barrier or quantum tunneling through time-dependent barriers become important issues [171] . The simplest account of dissipative terms and their influence on activated processes, i.e., the correction to, for instance, rates obtained using transition state theory (see Appendix A), comes about in Kramer's theory [172], which gives a correction factor to the transition state rate constant

$$\kappa = -\Delta + \sqrt{1 + \Delta^2} \tag{7.44}$$

where $\Delta = \eta/2\omega_b$. Here η is the frictional constant and ω_b the numerical value of the imaginary barrier frequency, i.e., it can be obtained from the second derivative of the potential at the barrier. We notice that κ is smaller than unity. For small values of the friction, κ approaches unity, and for large values, $\kappa \to \omega_b/\eta$. Thus the rate decreases with increasing friction, which can be understood by enhanced recrossing of the transition state caused by the frictional forces. On the other hand, it is known that tunneling of a particle through a barrier coupled to a heat bath of oscillators is enhanced with the excitation of these oscillators, i.e., with surface temperature [173]; compare the discussion of phonon-assisted tunneling in section 3.2. Thus there are two opposing effects and which one is dominating will depend on the surface temperature, with the frictional effect dominating at low surface temperatures. In this connection it could be mentioned that tunneling even through barriers with barrier heights of the order 0.5–2 eV appears to play a role for dissociative sticking probabilites of nitrogen [149, 155]. This shows that the barrier height is not the only important parameter, but that also the barrier width and shape as well as the coupling to the substrate motion are important factors.

EXERCISES

7.3: Derive eq. (7.35) and eq. (7.43).

7.4: Show that the fluctuating force defined by eq. (7.43) obeys the fluctuation-dissipation theorem with zero memory, i.e., that

$$\langle f(0)f(t)\rangle = k_B T_s \eta m \delta(t) \tag{7.45}$$

7.5: Use a Langevin equation of the type (7.41) without the force ω', i.e.,

$$m\frac{d}{dt}\mathbf{v}(t) = -m\eta\mathbf{v}(t) + \mathbf{f}(t) \tag{7.46}$$

and eq. (3.35) to obtain a connection between the surface diffusion constant and the friction coefficient η

$$D = \frac{k_B T}{m\eta} \tag{7.47}$$

The velocity $\mathbf{v}$ as well as the fluctuating force $\mathbf{f}$ have components in the x and y directions.

8

A SEMICLASSICAL APPROACH

Methods that are not either completely classical or quantum mechanical are denoted "semiclassical." In the present chapter we shall describe a semiclassical approach that is capable of reducing the numerical problem to one, that is more maneagable than that arising from a complete classical description, and that avoids introducing the partitioning in primary and secondary lattice atoms as described in the previous chapter. Furthermore, the theory works with a quantum boson description of the phonons.

Semiclassical treatments of atom/molecule surface dynamics have been considered by several groups [99, 100, 67, 104, 105]. These theories are very similar in spirit, but the method presented here has the advantage that it can in a relatively simple way account for surface temperature effects. The theory can be derived by assuming a Hartree product–type wavefunction

$$\psi = \Psi(X, Y, Z, t)\Phi(\{Q_k\}, t) \tag{8.1}$$

where X, Y, Z are the atoms and Q_k the normal mode coordinates for the solid. Here we just consider the atom-surface case, but the final equations can easily be extended to molecule-surface collisions. Inserting the trial function (8.1) in the time-dependent Schrödinger equation we get

$$i\hbar\frac{\partial\Psi}{\partial t} = \langle\Phi|H|\Phi\rangle\Psi - i\hbar\left\langle \Phi|\frac{\partial\Phi}{\partial t}\right\rangle\Psi \tag{8.2}$$

$$i\hbar\frac{\partial\Phi}{\partial t} = \langle\Psi|H|\Psi\rangle\Phi - i\hbar\left\langle \Psi|\frac{\partial\Psi}{\partial t}\right\rangle\Phi \tag{8.3}$$

where the brackets in the first equation indicate integration over phonon coordinates and in the second over atom coordinates. This treatment of the coupling between the two systems is expected to be sufficiently accurate if the coupling between them is weak [102].

The Hamiltonian operator can be divided as follows

$$H = H_m + H_0 + V_{int} \tag{8.4}$$

i.e., in a hamiltonian for the isolated systems for the atom H_m, the solid H_0, and the interaction potential V_{int}. Considering furthermore a linear expansion of the potential in the phonon coordinates we have

$$V_{int} = V_0 + \sum_k V_k^{(1)} Q_k \tag{8.5}$$

Thus we obtain the following equations for the phonons

$$i\hbar \frac{\partial \Phi}{\partial t} = \left(\langle \Psi | H_m | \Psi \rangle - i\hbar \left\langle \Psi | \frac{\partial \Psi}{\partial t} \right\rangle \right) \Phi + (H_0 + \langle \Psi | V_{int} | \Psi \rangle) \Phi \tag{8.6}$$

and for the atom

$$i\hbar \frac{\partial \Psi}{\partial t} = \left(E_{int} - i\hbar \left\langle \Phi | \frac{\partial \Phi}{\partial t} \right\rangle \right) \Psi + \left(H_m + V_0 + \sum_k \langle \Phi | Q_k | \Phi \rangle V_k^{(1)} \right) \Psi \tag{8.7}$$

where $E_{int} = \langle \Phi | H_0 | \Phi \rangle$. We now notice that both equations contain some phasefactors, i.e., terms which depend only upon time. These phasefactors can be transformed out of the equations by the following substitutions

$$\tilde{\Phi} = \Phi \exp \left(\frac{i}{\hbar} \left(\int dt \left(\langle \Psi | H_m | \Psi \rangle - i\hbar \left\langle \Psi | \frac{\partial \Psi}{\partial t} \right\rangle \right) \right) \right) \tag{8.8}$$

$$\tilde{\Psi} = \Psi \exp \left(\frac{i}{\hbar} \left(\int dt \left(E_{int} - i\hbar \left\langle \Phi | \frac{\partial \Phi}{\partial t} \right\rangle \right) \right) \right) \tag{8.9}$$

In the equation for the phonon wavefunction we shall furthermore replace the expectation value by the classical limit, i.e., $\langle \Psi | V_0 | \Psi \rangle \sim V_0(\{x_i(t), y_i(t), z_i(t)\})$, and so on. This approximation is consistent with the assumption of a narrow wavepacket for the atom (see below). Then the equation we solve for the phonons is identical to the one solved using classical mechanics for the atom motion. We then finally obtain the following equations

$$i\hbar \frac{\partial \Phi}{\partial t} = \left(H_0 + V_0(t) + \sum_k V_k^{(1)}(t) Q_k \right) \Phi \tag{8.10}$$

$$i\hbar \frac{\partial \Psi}{\partial t} = \left(H_m + V_0 + \sum_k \varepsilon_k(t) V_k^{(1)} \right) \Psi \tag{8.11}$$

where $\varepsilon_k(t) = \langle \Phi | Q_k | \Phi \rangle$ and where $\tilde{\Phi}$ and $\tilde{\Psi}$ have been replaced by Φ and Ψ.

The equation for the phonons is separable since

$$H_0 = \frac{1}{2} \sum_k (p_k^2 + \omega_k^2 Q_k^2) \tag{8.12}$$

and hence the wavefunction may be written as $\Phi = \Pi_{k=1}^{3N-6} \phi_k(Q_k)$, where N is the number of atoms in the solid. Thus eq. (8.10) can be transformed to $3N$-6 linearly forced oscillator equations. The solution of the forced oscillator problem is well known in the literature [103, 101] and will be considered in more detail below.

The method outlined above assumes a classical mechanical treatment of the motion of the gas-phase atoms but quantizises the lattice motion. If the lattice is treated within a harmonic approximation, it is advantageous from a computational point of view to introduce a quantum mechanical description, the reason being that quantum operator technique in this case gives a solvable algebra. Furthermore, we avoid the averaging over oscillator phase angles, which is necessary when using the classical description of the oscillators. Problems with the proper treatment of zero-point vibrational energy of the lattice are also avoided.

According to eq. (8.11) the gas atoms move in an average potential (a mean field potential) of the phonons. However, in order to obtain a temperature-dependent description we need to add some corrections to the mean field theory [106]. These corrections have to do with the correction for "detailed balance" and are well known from gas-phase molecular dynamics [109]. These corrections will be described in further detail below.

The classical limit for the motion of the incoming atom/molecule can be obtained from eq. (8.11) by using a Gaussian wavepacket (GWP) of the form

$$G(\mathbf{R}, t) = \exp\left(\frac{i}{\hbar}((\mathbf{R} - \mathbf{R}(t))^{+}\mathbf{A}(\mathbf{R} - \mathbf{R}(t)) + \mathbf{P}_R \cdot (\mathbf{R} - \mathbf{R}(t)) + \gamma(t)\right) \tag{8.13}$$

for Ψ in eq. (8.11). In (8.13) $\mathbf{R}(t) = (X(t), Y(t), Z(t))$, $\mathbf{P}_R(t)$, and $\gamma(t)$ are time-dependent parameters. If now the interaction potential is expanded to second order in $\Delta\mathbf{R} = (\mathbf{R} - \mathbf{R}(t))$ and the Gaussian wave packet (GWP) is inserted in the time-dependent schrödinger equation (TDSE) (8.11) we can, by equating terms of equal powers of $\Delta X = X - X(t)$, $\Delta X \Delta Y$, and so on. obtain the equations of motion for the time-dependent parameters [107]. In this manner we get from eq. (8.11), using $H_m = -(\hbar^2/2m)\nabla_R^2$, the following equations of motion

$$\dot{\mathbf{R}}(t) = \tfrac{1}{m}\mathbf{P}_R(t) \tag{8.14}$$

$$-\dot{\mathbf{P}}_R(t) = \nabla_R W|_{\mathbf{R}=\mathbf{R}(t)} \tag{8.15}$$

$$-\dot{\mathbf{A}}(t) = \tfrac{2}{m}\mathbf{A}^2 + \nabla_R^2 W|_{\mathbf{R}=\mathbf{R}(t)} \tag{8.16}$$

$$\dot{\gamma}(t) = \tfrac{i\hbar}{m} tr\mathbf{A} - W(\mathbf{R}(t)) + \frac{1}{2m}\mathbf{P}_R^2 \tag{8.17}$$

where $W(\mathbf{R}) = V_0 + \Sigma_k \varepsilon_k(t) V_k^{(1)}$. Thus the potential and the derivatives are evaluated at the center of the wavepacket, and this center moves according to classical equations of motion.

This method has also been used for obtaining semiclassical equations of motion for atom-surface scattering by Jackson [108]. If only the equations for $\mathbf{R}(t)$ and $\mathbf{P}_R(t)$ are used we obtain classical dynamical equations of motions for the translational motion of the atom moving in an Ehrenfest average potential, i.e., a potential which has been averaged over the quantum coordinates of the system (here the solid). Thus the equations of motion are solved in a mean field potential of the solid. This mean field potential will be considered below.

8.1 NORMAL MODE COORDINATES

The transformation between displacement and normal mode coordinates is given as (see also section 2.3)

$$\eta_{\alpha\gamma} = \sum_r T_{\alpha\gamma;r} Q_r \tag{8.18}$$

where $\eta_{\alpha\gamma}$ is the mass-weighted displacement from equilibrium of atom α in direction γ (see below). $\mathbf{T}$ is the matrix which diagonalizes the force constant matrix. In the harmonic approximation the hamiltonian for the solid is simply

$$H_c = V_0 + \sum_{r=1}^{M} H_r \tag{8.19}$$

where

$$H_r = \tfrac{1}{2}(\dot{Q}_r^2 + \omega_r^2 Q_r^2) \tag{8.20}$$

As mentioned the $M = 3N - 6$ nonzero frequencies are obtained by diagonalizing the second derivative matrix (the force constant matrix). Figure 8.1 shows $N(\omega)$, the frequency distribution for a small crystal consisting of 134 or 910 atoms. Quantities such as energy accommodation, scattering angles, surface temperature effects, and other quantities can be estimated by using a crystal size of a few hundred atoms.

We now quantize the vibrational modes r and write the normal mode coordinates in terms of the phonon creation/annihilation operators

$$Q_r = b_r(a_r^+ + a_r) \tag{8.21}$$

where $b_r = (\hbar/2\omega_r)^{1/2}$. The operator a_r^+ creates a quantum in mode r when operating on the phonon wavefunction, i.e.,

$$a_r^+|n_r\rangle = \sqrt{n_r + 1}|n_r + 1\rangle \tag{8.22}$$

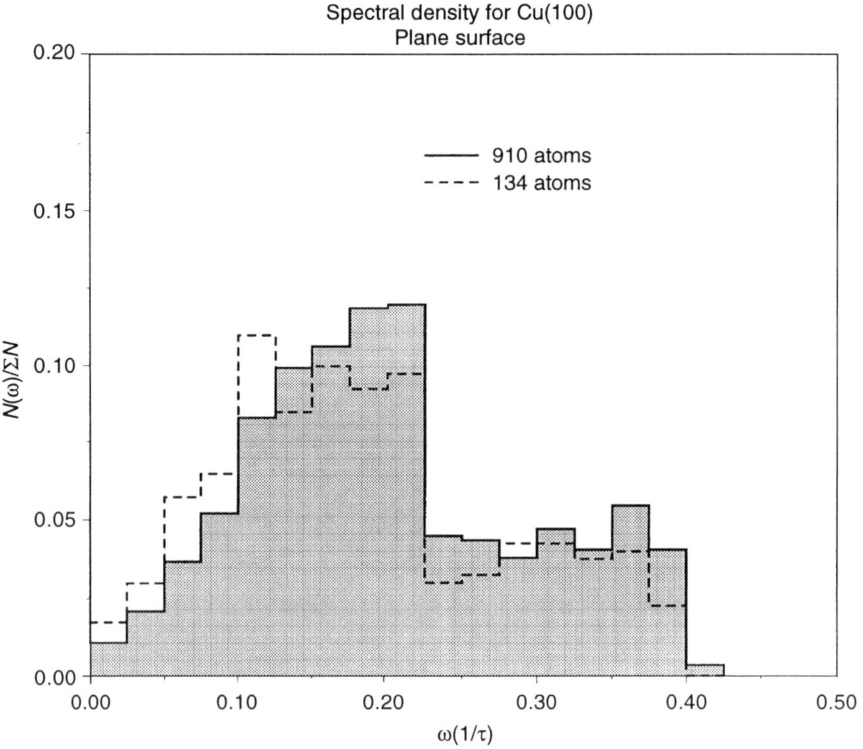

FIGURE 8.1 The phonon distribution $g(\omega) = \sum N(\omega)/N$ for a finite crystal as a function of ω in units of τ^{-1} where $\tau = 10^{-14}$ sec. The sum runs over phonon frequencies in a small $\Delta\omega$ range.

and for the a_r operator we have

$$a_r|n_r\rangle = \sqrt{n_r}|n_r - 1\rangle \tag{8.23}$$

The probability (P_{n_k}) for being in quantum state n_k for oscillator k is assumed to be given by a Boltzmann distribution, i.e.,

$$P_{n_k} = z_k^{n_k}(1 - z_k) \tag{8.24}$$

where

$$z_k = \exp(-\beta\hbar\omega_k) \tag{8.25}$$

and $\beta = 1/k_BT_s$. Transitions among the phonon quantum states may now be induced by anharmonic terms of the solid or through coupling to the electronic motion (electron-phonon coupling).

Transitions induced by gas-surface collisions are, however, the processes which we are concerned with in the present chapter. As mentioned in Chapter 4, the potential can be written as a sum of atom-atom pair potentials if we wish to extract the phonon coupling from it in an easy manner. But the EDIM potential also contains explicitly the important information on phonon forcing. Here we just assume that the interaction potential can be written as

$$V_I = \sum_{\alpha} V_{a\alpha}(R_{a\alpha}) \tag{8.26}$$

where $R_{a\alpha}$ is the distance from the gas-phase atom a to the surface atom α, i.e.,

$$R_{a\alpha} = \sqrt{\sum_{i=1,2,3} (\overline{X}_i - X_{\alpha i})^2} \tag{8.27}$$

where we have replaced X, Y, Z by $\overline{X}_i$ $(i = 1, 2, 3)$.

The interaction potential is now expanded in normal mode coordinates

$$V_I = V_I^{(0)} + \sum_r V_r^{(1)} Q_r + \frac{1}{2}\sum_{rr'} V_{rr'}^{(2)} Q_r Q_{r'} + \cdots \tag{8.28}$$

where $V_I^{(0)}$ is the interaction potential with the lattice atoms in their equilibrium position $X_{\alpha i}^{eq}$ and

$$V_r^{(1)} = \sum_{\alpha} \frac{\partial V_{a\alpha}}{\partial R_{a\alpha}} \frac{\partial R_{a\alpha}}{\partial Q_r}\bigg|_{eq} \tag{8.29}$$

The second derivative denoted by $V_{rr'}^{(2)}$ is given as

$$V_{rr'}^{(2)} = \sum_{\alpha} \frac{\partial V_{a\alpha}}{\partial R_{a\alpha}}\bigg|_{eq} \frac{\partial^2 R_{a\alpha}}{\partial Q_r \partial Q_{r'}} + \sum_{\alpha\beta} \frac{\partial^2 V_{a\alpha}}{\partial R_{a\alpha}\partial R_{a\beta}}\bigg|_{eq} \frac{\partial R_{a\alpha}}{\partial Q_r} \frac{\partial R_{a\beta}}{\partial Q_{r'}} \tag{8.30}$$

In order to calculate $(\partial R_{a\alpha}/\partial Q_r)$ we use eqs. (8.18) and (8.27) and the fact that

$$\eta_{\alpha\gamma} = m_{\alpha}^{1/2}(X_{\alpha\gamma} - X_{\alpha\gamma}^{eq}) \tag{8.31}$$

i.e.,

$$\frac{\partial R_{a\alpha}}{\partial Q_r} = -m_{\alpha}^{-1/2} \sum_i T_{\alpha i;r}(\overline{X}_i - X_{\alpha i}^{eq})/R_{a\alpha}^{eq} \tag{8.32}$$

The second derivatives can be found in a similar manner

$$\frac{\partial^2 R_{a\alpha}}{\partial Q_r \partial Q_{r'}} = \frac{1}{m_{\alpha} R_{a\alpha}^3} \sum_{ij} T_{\alpha i;r} T_{\alpha j;r'} (\overline{X}_i - X_{\alpha i}^{eq})(\overline{X}_j - X_{\alpha j}^{eq})(\delta_{ij} - 1) \tag{8.33}$$

The terms $V_r^{(1)}$ and $V_{rr'}^{(2)}$ in (8.28) are those responsible for collision-induced phonon excitation. These terms depend upon time through the time-dependence of the trajectory. If only the linear terms $V_r^{(1)}$ are included, the oscillators are called "linearly forced." From extensive studies on the linearly forced (and also more general) atom-molecule interactions, we know that good agreement between the semiclassical description and the full quantum description of the excitation process is obtained through the so-called "symmetrized Ehrenfest" approach. This approach is based upon the following two conjectures:

1. The classical particle motion should be governed by an average potential $V_{eff} = \langle\Psi|V_I|\Psi\rangle$, where Ψ is the total wavefunction of the quantum system, i.e., the phonons in the present case. The averaging is carried out over the quantum variables Q_k.
2. If the quantum system makes a transition from quantum state n to state n' with the probability $P_{n\to n'}$ as a result of the collision with the classical particle (with mass m), then one should calculate this probability using an average velocity $v_{av} = (v_n + v_{n'})/2$ where $v_n = \sqrt{2(E - E_n)/m}$, E the total energy, and E_n the energy of the quantum state n.

These "rules" were originally suggested so as to impose quantum detailed balance in the results obtained with the quantum-classical model for energy transfer in molecular collisions (for a review see [57]). Later it was found that the initial momentum could be obtained variationally [58], and that the result for the optimal variationally determined momentum was very close to the above simple arithmetic mean velocity. The symmetrization principle is, therefore, theoretically well founded and is well known in gas-phase collisions and has been used also in surface scattering [59, 101]. Thus for the forced oscillator problem this approach is very accurate [58] (see also Table 8.1). By introducing the normal mode oscillators in the theory of gas-surface collisions we are converting the problem to that of many forced "independent" oscillators. Hence the approach, which is accurate for a single forced oscillator, will also be accurate when used in this context.

8.2 A MODEL SYSTEM

In order to illustrate how the detailed balance correction works, let us consider the collision of an atom and a harmonic oscillator with a linear forcing from the interaction potential, i.e., we assume

$$V^{(1)} = f(R(t))Q = f(t)Q \tag{8.34}$$

The total hamiltonian is then

$$H = \hat{H}_0 + V^{(1)} + V^{(0)} + \frac{1}{2\mu}P_R^2 \tag{8.35}$$

TABLE 8.1 Transition Probabilities for an Atom Colliding with a Harmonic Oscillator

Number	Energy	Transition	Semi Classical	Exact	Number	Energy	Transition	Semi Classical
1	7.6	0–1	4.29(−5)	4.30(−5)	6	16.8365	0–1	0.357
	7.6	0–2	1.38(−11)	1.28(−11)		16.8365	0–2	0.119
	8.8	0–1	2.02(−4)	2.03(−4)		16.8365	1–2	0.323
	8.8	1–2	2.21(−5)	2.23(−5)		16.8365	2–3	0.308
	10.0	0–1	6.57(−4)	6.58(−4)		16.8365	2–4	0.147
	10.0	1–2	1.51(−4)	1.52(−4)		16.8365	2–5	0.016
	12.0	0–1	2.84(−3)	2.85(−3)	8	6.0	0–1	2.14(−2)
	12.0	1–2	1.31(−3)	1.32(−3)		8.0	0–1	9.76(−2)
	16.0	0–1	1.90(−2)	1.92(−2)		8.0	1–2	4.06(−2)
	16.0	1–2	1.64(−2)	1.66(−2)		10.0	0–1	0.211
	16.0	2–3	8.38(−3)	8.49(−3)		10.0	1–2	0.171
	20.0	0–1	6.21(−2)	6.39(−2)		10.0	2–3	5.81(−2)
	20.0	1–2	7.00(−2)	7.14(−2)		12.0	0–1	0.310
	20.0	2–3	5.42(−2)	5.50(−2)		12.0	1–2	0.308
6	12.8365	0–1	0.276	0.317		16.0	0–1	0.354
	12.8365	0–2	4.05(−2)	4.86(−2)		16.0	1–2	0.242
	12.8365	1–2	0.305	0.335		16.0	2–3	0.258
	12.8365	1–3	4.04(−2)	4.47(−2)		20.0	0–1	0.250
	14.8365	0–1	0.328	0.378		20.0	0–2	0.264
	14.8365	1–2	0.343	0.376		20.0	1–3	0.199
	14.8365	1–3	9.46(−2)	0.110		20.0	2–4	0.189

The semiclassical values are obtained using the symmetrized Ehrenfest approach. The energies are in units of $\frac{1}{2}\hbar\omega$. The three systems correspond to various reduced atom oscillator masses. System no. 1 has a relatively large oscillator mass, whereas systems no. 6 and 8 have small oscillator masses. The atoms of the surface are often much heavier than the gas atoms. Thus this case corresponds mainly to system no. 1. The numbering of the system refers to that of the first paper where the problem was solved exactly [60].
(−5) means 10^{-5}, etc.,

where $\hat{H}_0$ is the hamiltonian for the quantum oscillator

$$\hat{H}_0 = \hbar\omega\left(a^+a + \frac{1}{2}\right) \tag{8.36}$$

$V^{(0)}$ the potential for $Q = 0$, and the last term the kinetic energy of the atom. Asymptotically the total energy is

$$E = E_n + E_{kin} = \hbar\omega\left(n + \frac{1}{2}\right) + \frac{1}{2\mu}P_R^2(t_0) \tag{8.37}$$

where μ is the reduced mass of the atom-diatom system. The first conjecture now says that the relative (classical) motion is governed by the effective hamiltonian

$$H_{eff} = \langle\Psi|H|\Psi\rangle = \langle\Psi|\hat{H}_0|\Psi\rangle + \langle\Psi|V^{(1)}|\Psi\rangle + V^{(0)} + \frac{1}{2\mu}P_R^2 \tag{8.38}$$

where the two first terms will be evaluated below.

The second conjecture says that the initial velocity should be

$$\frac{1}{\mu}P_R(t_0) = \frac{1}{2}(v_n + v_m) \tag{8.39}$$

for the quantum transition n to m. The classical trajectory is obtained from the effective hamiltonian and is defined by the hamilton equations of motion, i.e.,

$$\dot{R} = \tfrac{1}{\mu}P_R \tag{8.40}$$

and

$$\dot{P}_R = -\frac{\partial}{\partial R}H_{eff} \tag{8.41}$$

This trajectory defines the time dependence of the perturbation on the quantum system, i.e., $f(R(t)) = f(t)$ in eq. (8.34). In order to obtain the probability $P_{n\to m}$ and the wavefunction to be used in (8.38), we should solve the time-dependent Schrödinger equation

$$i\hbar\frac{\partial\Psi}{\partial t} = H\Psi \tag{8.42}$$

Introducing the evolution operator such that

$$\Psi(t) = U(t, t_0)\Psi(t_0) \tag{8.43}$$

we get, in the interaction representation $U(t, t_0) = \exp(-iH_0t/\hbar)U_I(t, t_0)$, the following equation for U_I

$$i\hbar\frac{dU_I(t, t_0)}{dt} = \exp(iH_0t/\hbar)V^{(1)}(R(t))\exp(-iH_0t/\hbar)U_I(t, t_0) \tag{8.44}$$

Introducing the quantum operators a^+ and a^- we can write the equation as

$$\begin{aligned} i\hbar\frac{dU_I(t, t_0)}{dt} &= \exp\left(i\hbar\omega\left(a^+a + \frac{1}{2}\right)\right)bf(t)(a^+ + a) \\ &\quad\times\exp\left(-i\hbar\omega\left(a^+a + \frac{1}{2}\right)\right)U_I(t, t_0) \end{aligned} \tag{8.45}$$

where b is defined below. This equation is solvable analytically, and in the operator representation one obtains [156]

$$U_I(t, t_0) = \exp(i(\beta(t, t_0) + \alpha^+(t, t_0)a^+ + \alpha^-(t, t_0)a)) \tag{8.46}$$

where

$$\beta = \frac{1}{\hbar}\int_{t_0}^{t} dt' f(t')[\exp(i\omega t')\alpha^-(t') - \exp(-i\omega t')\alpha^+(t')] \tag{8.47}$$

and

$$\alpha^\pm = -\frac{b}{\hbar}\int_{t_0}^{t} dt' f(t')\exp(\pm i\omega t') \tag{8.48}$$

where b is a matrix element $b = \sqrt{\hbar/2\omega}$. The probability for a quantum transition from n to m is then

$$P_{nm} = |\langle m|U(t, t_0)|n\rangle|^2 \tag{8.49}$$

where

$$\exp(i\beta - \rho/2)(i\alpha^+)^{m-n}\sqrt{n!m!}\sum_{k=0}^{n}\frac{(-1)^k}{k!}\frac{\langle m|U(t, t_0)|n\rangle = \rho^k}{(m-n+k)!(n-k)!} \tag{8.50}$$

and

$$\rho = \alpha^+\alpha^- = |\alpha^+|^2 \tag{8.51}$$

From eq. (8.50) we see that the quantity ρ determines the magnitude of the transition probability, ρ is sometimes called the excitation strength, and for small values of ρ we have

$$P_{nm} \sim \rho^{|m-n|} \tag{8.52}$$

In order to determine the value of ρ and thereby the probability, one has to evaluate the integrals (8.48), i.e., the Fourier transform of the force $f(t)$ driving the oscillator.

The integral (8.48) may be evaluated analytically in a number of cases. In actual calculations it is of course calculated numerically, but for illustrative purposes we need an analytical estimate. In order to obtain this estimate we assume an interaction potential of the form (see Fig. 8.2)

$$V(z) = A\exp(-2\beta z) - B\exp(-\beta z) \tag{8.53}$$

where z is the distance from the surface. The hamiltonian for the motion along the z-axis is given by the equation

$$E_0 = \tfrac{1}{2}m\dot{z}^2 + V(z) = \tfrac{1}{2}m(v_0\cos\theta)^2 \tag{8.54}$$

where m is the mass of the incoming atom, θ the approach angle, and $v_0\cos\theta$ the velocity component in the z-direction. Using the fact that the distance z from the atom to the "infinite" mass of the crystal below the top layer is fixed, we get the force driving the oscillator as

$$f(t) = \beta(2A\exp(-2\beta z) - B\exp(-\beta z)) \tag{8.55}$$

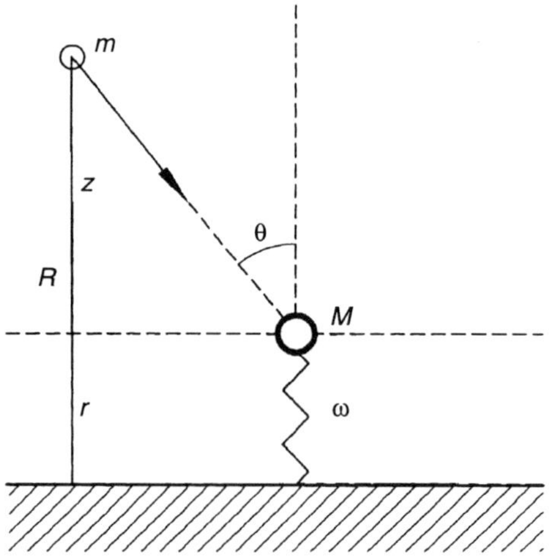

FIGURE 8.2 Schematic representation of the interaction of the atom with mass m with a "phonon" with frequency ω and mass M. Θ is the approach angle, z the distance to the surface, r the oscillator coordinate, and $R = z + r$.

where $z = z(t)$ can be obtained from the equation of motion (8.54). Thus we have (see exercise 8.1)

$$\exp(-\beta(z(t) - z_0)) = \frac{2}{(2 + b'y_0)\cosh(at) - b'y_0} \tag{8.56}$$

where z_0 is the turning point, i.e., where $\dot{z} = 0$, $y_0 = \exp(-\beta z_0)$, $b' = B/E_0$, and $a = \beta v_0 \cos\theta$. For this simple model we can evaluate the integral (8.48) as (see exercise 8.1)

$$\alpha^+ = -\frac{2\pi m\omega}{\beta\sqrt{2M\hbar\omega}} \frac{\cosh(\omega\tau/a)}{\sinh(\omega\pi/a)} \tag{8.57}$$

where τ is given by

$$\cos\tau = -\frac{b'y_0}{2 + b'y_0} \tag{8.58}$$

or equivalently

$$\cot\tau = -\sqrt{E_m/E_0} \tag{8.59}$$

where E_m is the well depth $E_m = B^2/4A$. We note that the mass M of the surface atom enters the matrix element, the reason being that normal modes have not been introduced at this point, i.e., we have not used mass-weighted coordinates.

In the perturbation limit, i.e., for small values of $|\alpha^+|^2 = \rho$, we have $\pi\omega/\beta v_0 \cos\theta > 1$ and hence we obtain from eq. (8.57)

$$\rho = C\exp\left(-\frac{2\pi\omega}{\alpha v_0\cos\theta}\left(1 - \frac{2\phi}{\pi}\right)\right) \tag{8.60}$$

where $C = 2\pi^2 m^2 \omega / \beta^2 \hbar M$ is independent of v_0, the initial velocity,

$$\phi = \tau - \pi/2 \tag{8.61}$$

and $\alpha = 2\beta$ is the slope parameter of the short-range repulsive potential.

According to the second conjecture above, we should replace v_0 with $(1/2)(v_n + v_{n'})$ when calculating the probability $P_{nn'}$. Inserting this in eq. (8.60), we obtain

$$\rho = C \exp\left(-\frac{2\pi\omega}{\alpha v_n \cos\theta}\left(1 - \frac{2\phi}{\pi}\right)\right) \exp(\mp|n' - n|x) \tag{8.62}$$

where

$$x = \frac{\pi\omega^2\hbar}{\alpha m v_n^3 \cos\theta}\left(1 - \frac{2\phi}{\pi}\right) \tag{8.63}$$

and where we have used the fact that

$$\tfrac{1}{2} m v_{n'}^2 = \tfrac{1}{2} m v_n^2 \mp |n' - n|\hbar\omega \tag{8.64}$$

where the upper sign is for excitation $(n' > n)$ and the lower for de-excitation $(n' < n)$. We have also assumed when deriving (8.62) that

$$|m - n|\frac{\hbar\omega}{m v_n^2} \ll 1 \tag{8.65}$$

This condition is fulfilled for thermal velocities and phonon frequencies less than 0.2–0.5 10^{14} sec^{-1}. These frequencies correspond to Debye temperatures of 2–400 K, i.e., the above expression is valid for the low-frequency modes, which also are those responsible for most of the energy accommodation at low collision energies.

Equation (8.62) shows that the symmetrization decreases the excitation and increases the de-excitation probabilities as compared to the original expression (8.60).

8.3 THE EFFECTIVE HAMILTONIAN

It is now relatively straightforward to extend this forced oscillator model for a single oscillator to the situation in a solid where we have a distribution of oscillators. Omitting the zero-point energy of the crystal we can write the total hamiltonian as

$$H = \frac{1}{2}\sum_{r=1}^{M}(P_r^2 + \omega_r^2 Q_r^2) + V_I^{(0)} + \sum_r V_r^{(1)} Q_r + \frac{1}{2}\sum_{rr'} V_{rr'}^{(2)} Q_r Q_{r'} + \frac{1}{2M} P_R^2 \tag{8.66}$$

where the first term is the energy of the solid, the second the static interaction potential with all solid atoms in their equilibrium positions, the first and second derivatives with respect to the phonon modes and finally the kinetic energy of the atom. We now quantize the bosons by introducing

$$Q_k = b_k(a_k^+ + a_k) \tag{8.67}$$

$$P_k = i\sqrt{\frac{\hbar\omega_k}{2}}(a_k^+ - a_k) \tag{8.68}$$

where $b_k = \sqrt{\hbar/2\omega_k}$. We then obtain the following hamiltonian expressed in terms of the boson operators

$$H = H_{cl} + \sum_k \hbar\omega_k \left(a_k^+ a_k + \frac{1}{2}\right) + \sum_k b_k V_k^{(1)}(a_k^+ + a_k)$$
$$+ \frac{1}{2}\sum_{kk'} V_{kk'}^{(2)} b_k b_{k'}(a_k a_{k'}^+ + a_k^+ a_{k'}^+) \tag{8.69}$$

where $H_{cl} = P_R^2/2M + V_I^{(0)}$. This term is treated classically and will therefore only contribute with an irrelevant phasefactor when we solve the TDSE for the bosons. It will therefore be dropped in the coming equations, but of course it will reappear in the effective hamiltonian.

Note that we have furthermore dropped the two-quantum operators $a_k a_{k'}$ and $a_k^+ a_{k'}^+$. Thus two quantum transitions will take place through successive operations with the one-quantum operators a_k and a_k^+. This approximation is introduced in order to obtain a solvable algebra for the operators. Such an algebra should be closed with respect to commutations. The above hamiltonian fulfills this condition.

We can once again switch to the interaction representation, i.e., we define

$$H_0 = \sum_k (\hbar\omega_k + V_{kk}^{(2)}) \left(a_k^+ a_k + \frac{1}{2}\right) \tag{8.70}$$

and

$$U = \exp\left(-i\int dt H_0/\hbar\right) U_I \tag{8.71}$$

The equation for U_I is then

$$i\hbar\frac{dU_I}{dt} = \left[\sum_k (f_k^+ a_k^+ + f_k^- a_k) + \sum_{k\neq k'} (f_{kk'}^+ a_{k'}^+ a_k + f_{kk'}^- a_{k'} a_k)\right] U_I \tag{8.72}$$

where $f_k^+ = V_k^{(1)}\exp(i\theta_k)$, $f_k^- = (f_k^+)^*$, $f_{kk'}^+ = \frac{1}{2}V_{kk'}^{(2)}\exp(i(\theta_{k'} - \theta_k))$, and $f_{kk'}^- = (f_{kk'}^+)^*$.

We have here introduced the modulated phonon frequency

$$\theta_k(t) = \int dt'(\omega_k + V_{kk}^{(2)}/\hbar) \tag{8.73}$$

where the last term depends upon time through the trajectory.

The evolution operator U_I expressed in terms of the boson operators is known for linearly and quadratically forced oscillators from semiclassical theories on energy transfer in polyatomic molecules [61]. Thus we have [61, 156]

$$U_I = U_{VV} U_{VT} \tag{8.74}$$

where $U_{VT} = \Pi_{k=1}^{M} U_{VT}^{(k)}$. The operators $U_{VT}^{(k)}$ satisfy the equation

$$i\hbar \frac{dU_{VT}^{(k)}}{dt} = U_{VV}^{+} \left(f_k^+ a_k^+ + f_k^- a_k^- \right) U_{VV} U_{VT}^{(k)} \tag{8.75}$$

where U_{VV} is a solution to

$$i\hbar \frac{dU_{VV}}{dt} = \sum_{k \neq l} \left(f_{kl}^+ a_k^+ a_l + f_{kl}^- a_k a_l^+ \right) U_{VV} \tag{8.76}$$

Both equations can be solved formally and expressed in terms of the operators. The solution to the VT problem has already been given in the previous section (see eq. (8.46)). We just have to replace $\alpha^{\pm}$ and β by

$$\alpha_k^{\pm} = -\frac{b_k}{\hbar} \int_{t_0}^{t} dt' f_k^{\pm}(t') \tag{8.77}$$

and

$$\beta_k = \frac{b_k}{\hbar} \int_{t_0}^{t} dt'(f_k^+(t')\alpha_k^-(t') - f_k^-(t')\alpha_k^+(t')) \tag{8.78}$$

For the VV problem defined by eq. (8.76) we have a solution of the form

$$U_{VV} = \Pi_{kl} \exp(\alpha_{kl}(t, t_0) a_k^+ a_l) \tag{8.79}$$

where the functions $\alpha_{kl}(t, t_0)$ are related to the solution of the equation

$$i\hbar \frac{d\mathbf{R}(t, t_0)}{dt} = \mathbf{B}\mathbf{R}(t, t_0) \tag{8.80}$$

where $\mathbf{R}(t_0, t_0) = \mathbf{I}$ (the unit matrix) and the elements of the $\mathbf{B}$ matrix are $B_{kl} = f_{kl}^+$ for $(k > l)$ and $B_{kl} = f_{kl}^-$ for $(k < l)$. At this point it is also convenient to define the matrix $\mathbf{Q} = \mathbf{R} - \mathbf{I}$. The functions α_{kl} can be expressed in terms of Q_{kl} [61]. If the second derivatives of the interaction potential are dropped, then $\mathbf{B} = 0$, $\mathbf{Q} = 0$, and then also $\alpha_{kl} = 0$, i.e., U_{VV} is the identity operator. In Appendix D we have given the connection between the Q_{kl} and the α_{kl} functions.

In order to evaluate the effective hamiltonian we can also use operator algebra, i.e., use the fact that

$$U_I^+ a_k U_I = a_k + i\alpha_k^+ + \sum_l Q_{kl}(a_l + i\alpha_l^+) \tag{8.81}$$

$$U_I^+ a_k^+ U_I = a_k^+ - i\alpha_k^- + \sum_l Q_{kl}^*(a_l^+ - i\alpha_l^-) \tag{8.82}$$

Thus the effective hamiltonian is obtained as

$$H_{eff} = \langle\Psi|H|\Psi\rangle = \langle\Psi_0|U^+ H U|\Psi_0\rangle \tag{8.83}$$

where $U(t, t_0)$ is the evolution operator (8.71) and $|\Psi_0\rangle$ the initial state, i.e.,

$$|\Psi_0\rangle = \Pi_{k=1}^M |n_k^0\rangle \tag{8.84}$$

The final expression for the effective hamiltonian is, after a little manipulation,

$$\begin{aligned} H_{eff} = {} & \sum_k \hbar\omega_k \left(n_k^0 + \frac{1}{2}\right) + E_{int} + V_I^{(0)} + \sum_k \omega_k^{-1} \varepsilon_k(t) \\ & + \frac{P_R^2}{2M} + \text{higher-order terms} \end{aligned} \tag{8.85}$$

where the first term is the initial energy of the oscillators, the second the energy transfer to the solid

$$E_{int} = \sum_k \hbar\omega_k \rho_k \tag{8.86}$$

with $\rho_k = \alpha_k^+ \alpha_k^-$. The next term is an effective potential

$$\varepsilon_k(t) = V_k^{(1)}(R(t)) \int dt' V_k^{(1)}(R(t')) \sin(\theta_k(t') - \theta_k(t)) \tag{8.87}$$

and the higher-order terms are

$$\begin{aligned} & \sum_k 2\mathrm{Re}Q_{kk} + \alpha_k^- \sum_j Q_{kj}\alpha_j^+ + \alpha_k^+ \sum_j Q_{kj}^* \alpha_j^- + \sum_{jj'} Q_{kj}^* Q_{kj'} \left(\delta_{jj'} n_j^0 + \alpha_j^- \alpha_{j'}^+\right) \\ & + \sum_k i b_k \left(f_k^- \sum_j Q_{kj}\alpha_j^+ - f_k^+ \sum_j Q_{kj}^* \alpha_j^- \right) \end{aligned}$$

$$+\frac{1}{2}\sum_{k\neq k'} V^{(2)}_{kk'} b_k b_{k'} \left[f^+_{kk'} \left(Q_{kk'} n^0_{k'} + Q_{k'k} n^0_k + \alpha^-_{k'}\alpha^+_k + \alpha^-_{k'} \sum_j Q_{kj}\alpha^+_j \right.\right.$$

$$\left. + \sum_j Q^*_{k'j} Q_{kj} n^0_j + \alpha^+_k \sum_j Q^*_{k'j}\alpha^-_j + \sum_{jj'} Q^*_{k'j} Q_{kj'} \alpha^-_j \alpha^+_{j'} \right)$$

$$+ f^-_{kk'} \left(Q_{k'k} n^0_k + \alpha^-_k \alpha^+_{k'} + Q^*_{kk'} n^0_{k'} + \alpha^-_k \sum_j Q_{k'j}\alpha^+_j + \sum_j Q^*_{kj} Q_{k'j} n^0_j \right.$$

$$\left.\left. + \alpha^+_{k'} \sum_j Q^*_{kj}\alpha^-_j + \sum_{jj'} Q^*_{kj} Q_{k'j'} \alpha^-_j \alpha^+_{j'} \right) \right] \qquad (8.88)$$

The functions $Q_{kl}(t)$ are given to first order by the simple expression

$$Q_{kl} = \frac{b_k b_l}{i\hbar} \int_{t_0}^{t} dt V^{(2)}_{kl} F^-_{kl}(t) \qquad (8.89)$$

but they are exactly obtainable from the solution to eqs. (8.80).

Furthermore, if ρ_k denotes the excitation strength for oscillator k, we have initially $\rho_k(t_0) = 0$ and finally for $t_1 \to \infty$ where the atom has left the surface, the only surviving terms in (8.85) are the initial energy, the kinetic energy, and E_{int}, such that

$$E_{int} = \frac{1}{2M}(P^2_R(t_0) - P^2_R(t)) \qquad (8.90)$$

for $t \to \infty$. Thus the energy transfer to the phonons E_{int} balances the energy lost by the incoming atom.

The term $\Sigma_k \varepsilon_k/\omega_k$ arises from the expectation value of the linear $V^{(1)}_r Q_r$ terms in (8.66). These terms are called phonon VT terms because they exchange one quantum of vibrational energy when forced by the translational motion. The second derivative terms in (8.66) contain operators such as $a^+_r a_{r'}$ and $a_r a^+_{r'}$, i.e., these terms create a quantum in one mode and destroy one in another. It has, however, turned out that these so-called VV terms are of minor importance for energy transfer in solids [62]. These terms may therefore be neglected. Thus if the higher-order terms are dropped we also have the simplification $\theta_k(t) = \omega_k t$ in eq. (8.73).

Eq. (8.86) shows that $E_{int} > 0$ irrespective of the initial state. However, the symmetrization principle discussed above showed that ρ_k should be replaced by

$$\rho_k \to \rho_k \exp(\mp|m-n|x_k) \qquad (8.91)$$

where x_k is given by eq. (8.63) with ω replaced by ω_k. In order to invoke this symmetrization, it is necessary to introduce an alternative derivation which divides the energy transfer into a phonon creation and a phonon annihilation process and then introduces the correction (8.91). If, as suggested above, we neglect the VV processes, we would get the following effective hamiltonian from eq. (8.85)

$$H_{eff} = E_{int} + V_I^{(0)}(R) + \sum_k \omega_k^{-1}\varepsilon_k(R,t) + \frac{P_R^2}{2M} \tag{8.92}$$

where we have dropped the initial energy of the solid. The equations of motion are obtained using Hamilton's principle, i.e.,

$$\frac{dH_{eff}}{dt} = \frac{\partial H_{eff}}{\partial t} + \frac{\partial H_{eff}}{\partial R}\dot{R} + \frac{\partial H_{eff}}{\partial P_R}\dot{P}_R \tag{8.93}$$

yields

$$\dot{R} = \frac{\partial H_{eff}}{\partial P_R} = \frac{P_R}{M} \tag{8.94}$$

and

$$\dot{P}_R = -\frac{\partial H_{eff}}{\partial R} = -\frac{\partial V_I^{(0)}}{\partial R} + \sum_k \omega_k^{-1}\frac{\partial \varepsilon_k(R,t)}{\partial R} \tag{8.95}$$

Thus if we require energy conservation $\dot{H}_{eff} = 0$ we get

$$\frac{\partial H_{eff}}{\partial t} = \sum_{k=1}^{M} \hbar\omega_k\dot{\rho}_k + \sum_k \omega_k^{-1}\frac{\partial \varepsilon_k}{\partial t} = 0 \tag{8.96}$$

or

$$\frac{\partial \varepsilon_k}{\partial t} = -\hbar\omega_k^2\dot{\rho}_k = -\omega_k\frac{d}{dt}\Delta E_k \tag{8.97}$$

where $\Delta E_k = \hbar\omega_k\rho_k$. Equation (8.97) may be used to evaluate the effective potential

$$\langle V^{(1)}\rangle = \sum_k \omega_k^{-1}\varepsilon_k(R,t) \tag{8.98}$$

from ΔE_k. Thus if the symmetrization is included in ΔE_k we can use eq. (8.97) to calculate the "symmetrized" potential. If a distribution of phonon states is introduced this potential will depend on the surface temperature. Thus a hot surface will be able to excite the molecule (see Table (8.2) and Fig. 8.3). The surface temperature dependence comes about only if x_k (8.63)

$$x_k = a\omega_k^2 \tag{8.99}$$

is nonzero. The parameter a can, as we have seen, be estimated from a simple model system.

TABLE 8.2 Initial Kinetic Energy $\langle E_{kin} \rangle$ and Final Kinetic Energy $\langle E'_{kin} \rangle$ in K as a Function of Surface Temperature T_s for Ar Atoms Hitting a Tungsten Surface

T_s	$\langle E_{kin} \rangle$	$\langle E'_{kin} \rangle$
393	326	539
590	327	575
1233	325	892
1785	330	972
349	1037	1022
1184	1116	1398
1757	1097	1650
331	2148	1924
609	2170	2057
840	2124	2143

The incident angle is $\theta = 45°$. Data from ref. [63]. The energy transfer to a hot surface is negative.

This derivation of a shows that it depends upon the interaction potential and the kinetic energy of the incoming atom/molecule. However, an alternative method for determining a without resorting to the model system would be to calculate the energy accommodation $\Delta E_{int}(T, T_s, a)$ such that we for $T = T_s$ have $\Delta E_{int}(T_s, T_s, a) = 0$, i.e., no energy accommodation if the gas temperature (T) and surface temperature (T_s) are the same. Below we see how the surface temperature can be built into the effective potential governing the motion of the atom. In order to do so in a simple manner, we first consider the energy transfer to the surface phonons. The trick is first to calculate ΔE_{int} and then use eqs. (8.97) and (8.98) to determine the potential which governs the motion of the atom.

8.3.1 Calculation of ΔE_{int}

We assume that the surface phonons initially are distributed according to some distribution function $p_{n_k^0}$. For a Boltzmann distribution we would have

$$p_{n_k^0}(T_s) = (1 - z_k) z_k^{n_k^0} \tag{8.100}$$

where $z_k = \exp(-\hbar\omega_k / k_B T_s)$. We furthermore use the fact that the energy transfer to the solid is small compared to the total energy of the solid, such that we can assume a constant temperature in the distribution function. This does not mean, however, that the atoms locally cannot have a higher temperature than the one assumed in the Boltzmann distribution function.

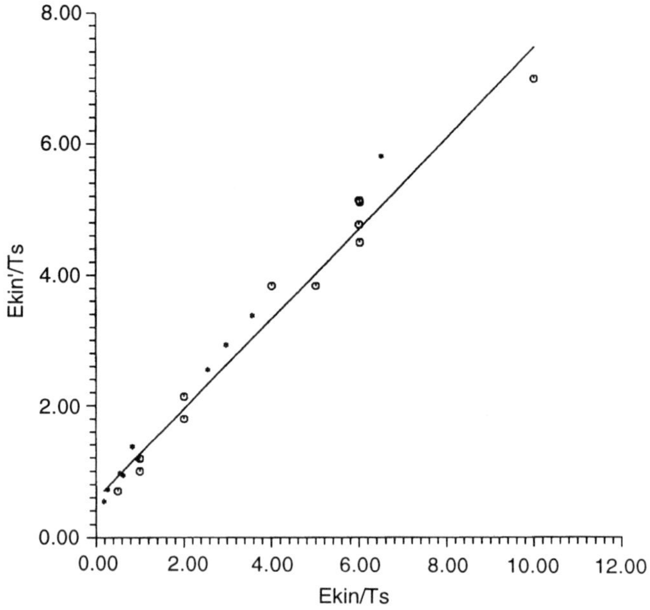

FIGURE 8.3 Initial (Ekin) and final (Ekin′) kinetic energy of Ar atoms hitting a tungsten surface scaled with the surface temperature (Ts). Stars denote data from [63] and circles calculations from [62]. We notice that at high surface temperatures the final kinetic energy is larger than the initial whereas the opposite is true for low surface temperature. The solid line is a least squares fit to the theoretical data.

The energy transfer to the solid is given by

$$\Delta E_{int} = \sum_k \sum_{n_k} \sum_{n_k^0} p_{n_k^0} (E_{n_k} - E_{n_k^0}) P_{n_k^0 \to n_k} \tag{8.101}$$

where $P_{n_k^0 \to n_k}$ is the probability for a transition from quantum state n_k^0 to n_k in mode k. Within the harmonic approximation we furthermore have

$$E_{n_k} - E_{n_k^0} = \hbar\omega_k \left(n_k - n_k^0\right) \tag{8.102}$$

We now use the fact that the excitation probability is (see eq. (8.50))

$$P_{n_k^0 \to n_k} = \exp(-\rho_k) \frac{n_k^0!}{n_k!} \rho_k^{n_k - n_k^0} \left[L_{n_k^0}^{n_k - n_k^0}(\rho_k) \right]^2 \tag{8.103}$$

where L_n^m is a Laquerre polynomial. For de-excitation we have likewise

$$P_{n_k^0 \to n_k} = \exp(-\rho_k) \frac{n_k!}{n_k^0!} \rho_k^{n_k^0 - n_k} \left[L_{n_k}^{n_k^0 - n_k}(\rho_k) \right]^2 \tag{8.104}$$

Inserting these expressions in (8.101) we get for excitation (+) and de-excitation (−) processes the energy transfer

$$\Delta E^{\pm}_{int} = \pm \sum_k \hbar\omega_k \exp(\rho_k(1+z_k)/(z_k-1))T^{\pm}_k \tag{8.105}$$

where

$$T^{\pm}_k = \sum_{m=0}^{\infty} mI_m(t_k)z_k^{\mp m/2} \tag{8.106}$$

and

$$t_k = 2\rho_k\sqrt{z_k}/(1-z_k) \tag{8.107}$$

where I_m is a Bessel function. The summations (8.105) and (8.106) may be carried out to give

$$\Delta E_{int} = \Delta E^{+}_{int} + \Delta E^{-}_{int} = \sum_k \hbar\omega_k\rho_k \tag{8.108}$$

The symmetrization is now introduced by using ρ^{+}_k in ΔE^{+}_{int} and ρ^{-}_k in ΔE^{-}_{int}, where

$$\rho^{\pm}_k = \rho_k \exp(\mp x_k m) \tag{8.109}$$

Here m is the number of quanta in the excitation/de-excitation process. The correction is introduced in the terms $T^{\pm}_k$, i.e., we use in (8.106)

$$t_k = 2\rho_k z_k^{1/2} \exp(\mp x_k m)/(1-z_k) \tag{8.110}$$

Once $\Delta E^{\pm}_{int}$ is obtained, we return to eq. (8.97) in order to obtain the potential connected to a certain energy transfer. Thus we write (see eq. (8.87))

$$\varepsilon_k(R,t) = V^{(1)}_k(R)\eta_k(t)\omega_k \tag{8.111}$$

where $\eta_k(t)$ is to be determined. From eq. (8.97) we get

$$V^{(1)}_k(R)\dot{\eta}_k(t) = -\frac{d\Delta E_k}{d\rho_k}\dot{\rho}_k \tag{8.112}$$

where

$$\Delta E_k = \Delta E^{+}_k + \Delta E^{-}_k \tag{8.113}$$

$$\Delta E^{\pm}_k = \pm\hbar\omega_k \exp[-\rho_k(1+z_k)/(1-z_k)]T^{\pm}_k \tag{8.114}$$

and where by using eq. (8.48) and (8.51), we get

$$\dot{\rho}_k = \frac{1}{\hbar\omega_k}(I_{sk}\sin(\omega_k t) + I_{ck}\cos(\omega t))V_k^{(1)}(R) \tag{8.115}$$

Inserting this in eq. (8.112) we have

$$\eta_k(t, T_s) = -\int dt'(\hbar\omega_k)^{-1}\frac{d}{d\rho_k}[\Delta E_k^+(T_s) + \Delta E_k^-(T_s)] \\ \times \{I_{sk}(t')\sin(\omega_k t') + I_{ck}(t')\cos(\omega_k t')\} \tag{8.116}$$

where

$$I_{ck}(t) = \int_{t_0}^{t} dt' V_k^{(1)}(R(t'))\cos(\omega_k t') \tag{8.117}$$

and

$$I_{sk}(t) = \int_{t_0}^{t} dt' V_k^{(1)}(R(t'))\sin(\omega_k t') \tag{8.118}$$

The effective time- and temperature-dependent potential is, in this model,

$$V_{eff} = V_I^{(0)}(R) + \sum_k \eta_k(t, T_s)V_k^{(1)}(R) \tag{8.119}$$

In the limit of $x_k = 0$ this expression is identical to the previously defined potential in eq. (8.92).

Although the approach given above may seem complicated, it is actually very simple to implement in practical calculations. The approach has many advantageous features:

- The surface temperature is built into the potential.
- The classical trajectory is not tied to a specific quantum transition. The velocity symmetrization is included when defining the effective potential.
- The time-consuming part, i.e., the evaluation of the effective potential, increases only linearly with the number of crystal atoms (N).
- One avoids treating some surface atoms in a special way (primary atoms)—all atoms are treated evenhandedly.
- We avoid averaging over oscillator phase angles. Only the aiming point has to be chosen randomly. Thus quantities such as energy accommodation converge rapidly (20–50 trajectories).
- If necessary, a phonon excitation spectrum is also obtained, i.e., information upon the most excited modes can be detected.

The expressions may be simplified further if the so-called "stochastic" approximation to the transition probabilities for excitation of the harmonic oscillators is introduced. Here the probabilities are given by the expression [156]

$$P_{n^0 \to n} = \frac{1}{\rho} \exp(-(n + n^0 + 1)/\rho) I_0 \left(\frac{2}{\rho} \sqrt{\left(n + \frac{1}{2}\right)\left(n^0 + \frac{1}{2}\right)} \right) \tag{8.120}$$

With this expression, we obtain the energy transfer to phonon mode k as

$$\Delta E_k = \hbar\omega_k \rho_k \frac{\sinh(\Delta_k)}{\Delta_k} \tag{8.121}$$

where $\Delta_k = \frac{1}{2}\beta\hbar\omega_k$ and $\beta = 1/k_B T_s$. This expression for the energy transfer is identical to the "exact" for small values of Δ_k. Using the stochastic model for the probabilities, we obtain the following expressions for the energy transfer to phonon mode k, divided in phonon creation and annihilation processes

$$\Delta E_k^{\pm} = \pm \frac{4\beta(\hbar\omega_k)^2}{\rho_k \sqrt{a_k}(\beta\hbar\omega_k \mp \sqrt{a_k})^2} \frac{\sinh(\Delta_k)}{\Delta_k} \tag{8.122}$$

where

$$a_k = \beta^2\hbar^2\omega_k^2 + 4\beta\hbar\omega_k\rho_k^{-1} \tag{8.123}$$

By substituting $\rho_k^{\pm}$ into the above equations we can obtain the temperature dependence of the energy transfer to the phonon mode. By omitting the last factor, $\sinh(\Delta_k)/\Delta_k$, we furthermore obtain an expression which behaves correctly in the limit $T_s \to 0$.

Table 8.3 shows results obtained with the semiclassical model described in this section. The results of 8 randomly selected trajectories are shown. The random

TABLE 8.3 Scattering of Hydrogen from a Cu(001) Surface

			$\theta = 0$					$\theta = 45$			
ΔX Å	ΔY Å	v'	j'	θ_f	ϕ_f	ΔE_{int}	v'	j'	θ_f	ϕ_f	ΔE_{int}
−1.81	0.64	0.97	1.50	12.9	268	0.48	0.97	1.61	10.6	181	0.32
−0.04	−0.46	0.96	1.61	10.9	251	0.82	0.98	1.30	29.0	175	0.22
1.25	1.16	0.98	1.37	11.5	216	0.70	0.98	1.49	28.5	167	0.19
−0.88	−1.59	1.00	1.45	12.8	−7.9	0.60	1.03	1.55	18.6	166	0.42
1.55	1.80	1.02	1.48	8.2	183.0	0.90	1.03	1.61	34.5	178	0.16
1.54	−0.41	1.05	1.04	10.9	59.0	0.42	1.04	1.57	14.1	187	0.41
0.88	1.18	0.97	2.05	7.7	156.0	0.55	0.98	1.02	18.0	157	0.26
0.93	0.73	0.97	1.60	3.7	−68.0	0.58	0.98	1.58	25.9	159	0.30

The approach angles are $(\theta, \phi) = (0, 0)$ and $(45°, 0)$. The initial vibrational/rotational states are $v = 1$ and $j = 1$. Final vibrational and rotational "states" are v' and j'. Surface temperature 300 K and initial kinetic energy 7.50 kJ/mol. The scattering angles are θ_f and ϕ_f in degrees. The energy transfer to the surface phonons is indicated by ΔE_{int} in kJ/mol.

variables are besides the aiming point ΔX, ΔY, the orientation of the molecule, the bond length, and an angle specifying the orientation of the angular momentum vector (see Chapter 6). Table 8.3 shows that the scattering from the Cu-surface is random in ϕ_f if the approach angle θ is zero. This is due to the surface corrugation. For $\theta = 45°$ the molecule is scattered mainly in the direction of $\phi_f \sim 180°$. The energy transfer to the surface amounts to about 5–10% of the initial kinetic energy. Little vibrational relaxation and rotational excitation is seen at this energy. The energy transfer decreases, as expected, with the smaller θ angle due to the smaller component of the velocity normal to the surface.
Figures 8.4 and 8.5 show, as a function of phonon frequency, the contributions to ΔE_k^+ and ΔE_k^- at $T_s = 300$ K and 800 K for He colliding with a Cu(001) surface. The values of x_k have been estimated using the expression (8.63). The kinetic energy is 10.9 kJ/mol and the incident angle $\theta = 45°$. We notice that at

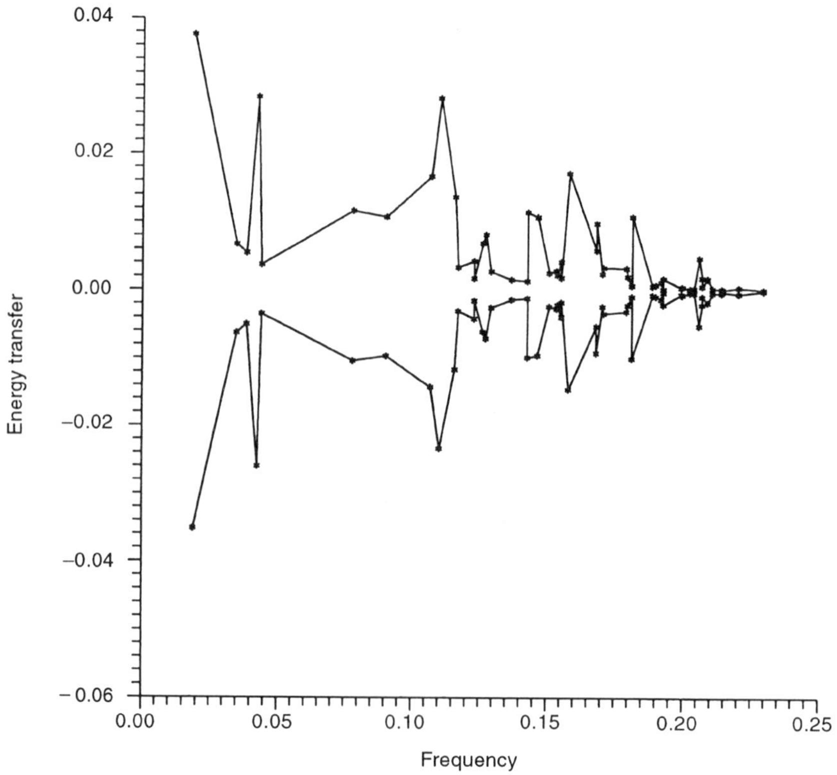

FIGURE 8.4 Energy transfer ΔE^+ (positive) and ΔE^- (negative) in units of kJ/mol for Heatoms colliding with a Cu(001) surface consisting of 244 atoms, as a function of phonon frequency (in units of 10^{-14} sec.). The surface temperature is 300 K, the initial kinetic energy is 10.9 kJ/mol, and incident angles $\theta = 45°$ and $\phi = 0°$. The lower panel shows the net energy transfer to the phonons. The numbers are obtained as average over 10 randomly selected trajectories. Only those frequencies which participate in the energy transfer are included.

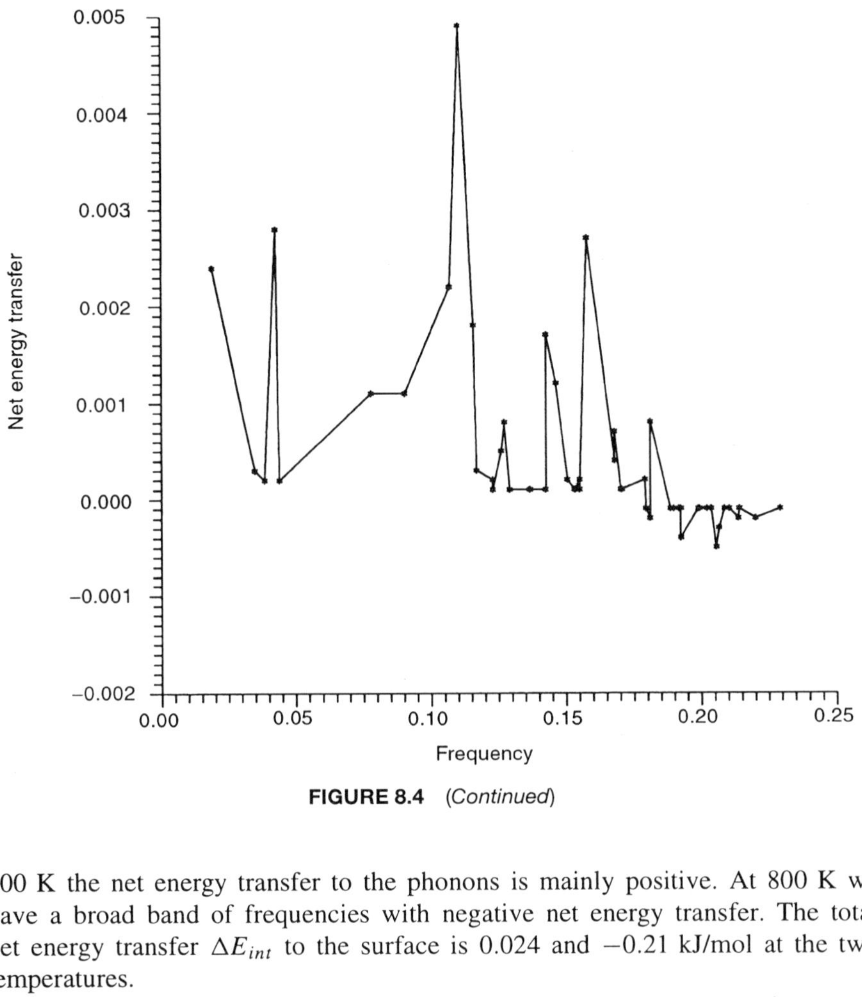

FIGURE 8.4 *(Continued)*

300 K the net energy transfer to the phonons is mainly positive. At 800 K we have a broad band of frequencies with negative net energy transfer. The total net energy transfer ΔE_{int} to the surface is 0.024 and −0.21 kJ/mol at the two temperatures.

Figure 8.6 shows the trajectory for a collision between hydrogen and a copper Cu(001) surface. For light masses, the phonon coupling may be neglected. It amounts to a few percent of the initial kinetic energy. We notice that the phonon energy transfer has a nonmonotonic behavior, showing that the surface temporarily may transfer energy to the molecule.

In the derivations of this chapter, we have assumed a collision of an atom and a solid consisting of a finite number of atoms. The approach is easily extended to treating any number of atoms in the gas phase using classical mechanics. However, the derivation was based upon the assumption of a weak coupling between the gas-phase atom(s) and the solid. This assumption makes the simple Hartree trial function a good approximation. If reaction between the gas-phase atoms and the solid atoms can occur, this separability approximation obviously breaks down. Thus, if surface etching or laser evaporation or other processes in which the surface atoms are removed from the surface are considered, the

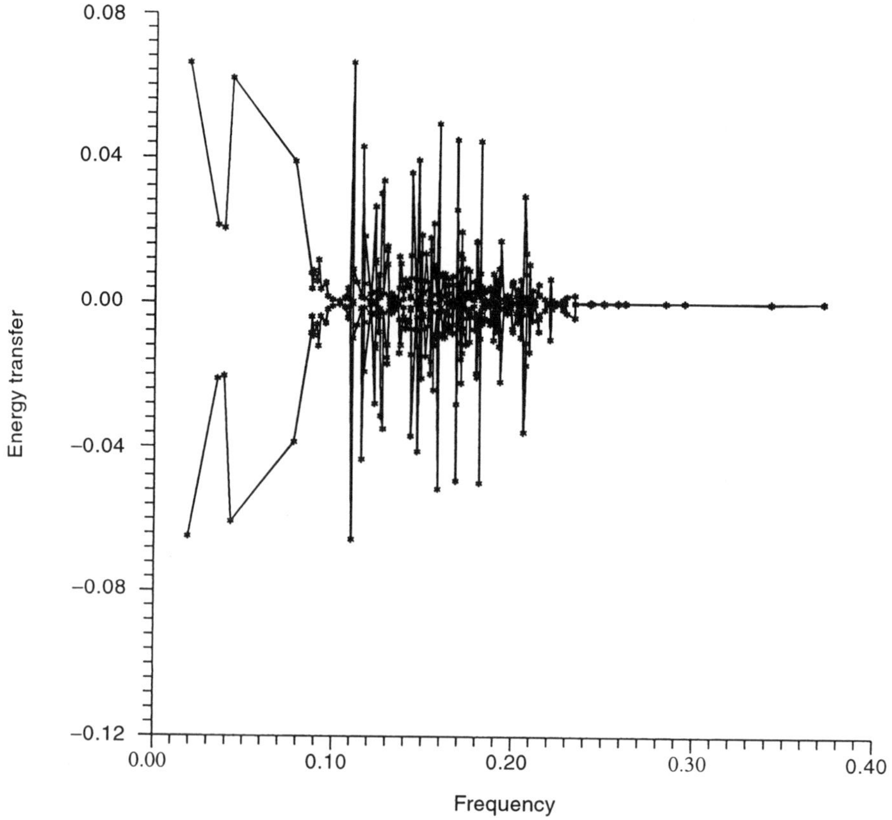

FIGURE 8.5 Same as figure 8.4 except that the surface temperature is 800 K. The same trajectories as those at 300 K were used.

dynamics of these atoms should be incorporated in the classical mechanical description on the same dynamical footing as the gas atoms. This can be accomplished by dividing the surface atoms into active and inactive ones, where the active are those which may react and the inactive the remaining. Thus a partitioning is made in much the same way as the one introduced when discussing the generalized Langevin equation. The inactive ones can then be treated using the quantum forced oscillator model as described in this section. This approach has, for instance, been used to study laser evaporation of sodium chloride [64].

EXERCISES

8.1: Derive expression (8.57) and the expression for $z(t)$.

8.2: Derive expression (8.121) by changing the summation to integration, using the fact that

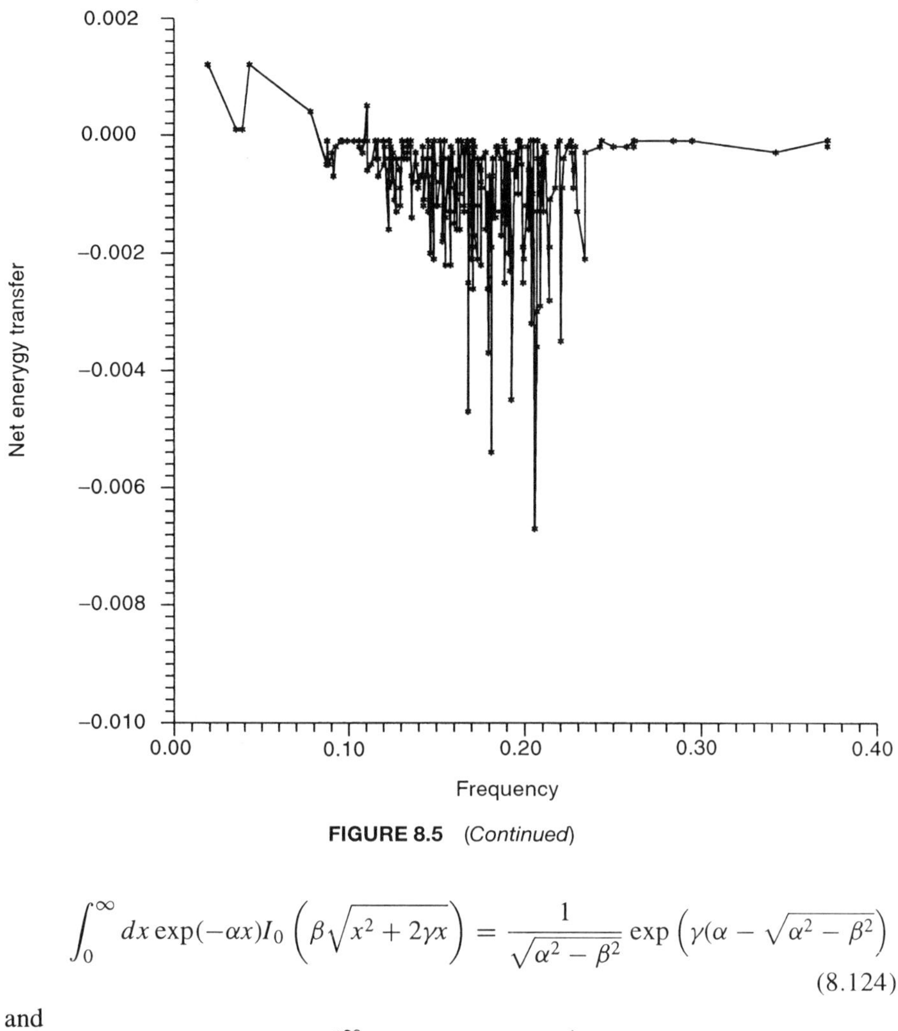

FIGURE 8.5 *(Continued)*

$$\int_0^\infty dx \exp(-\alpha x) I_0 \left(\beta\sqrt{x^2 + 2\gamma x}\right) = \frac{1}{\sqrt{\alpha^2 - \beta^2}} \exp\left(\gamma(\alpha - \sqrt{\alpha^2 - \beta^2}\right) \tag{8.124}$$

and

$$\int_0^\infty dxx \exp(-ax) = \frac{1}{a^2} \tag{8.125}$$

8.4 THE DEBYE-WALLER FACTOR

The phonon excitation can be probed by scattering with neutrons or He-atoms which have De Broglie wavelengths comparable to those of the phonons. For the probing of the "softer" surface phonons, scattering with low-energy He atom beams is ideal. The elastic scattering probabilities, i.e., the diffraction peaks, contain two factors. One depends on the structure of the solid while the other involves the temperature-dependent effects. Thus the fact that the experiments are

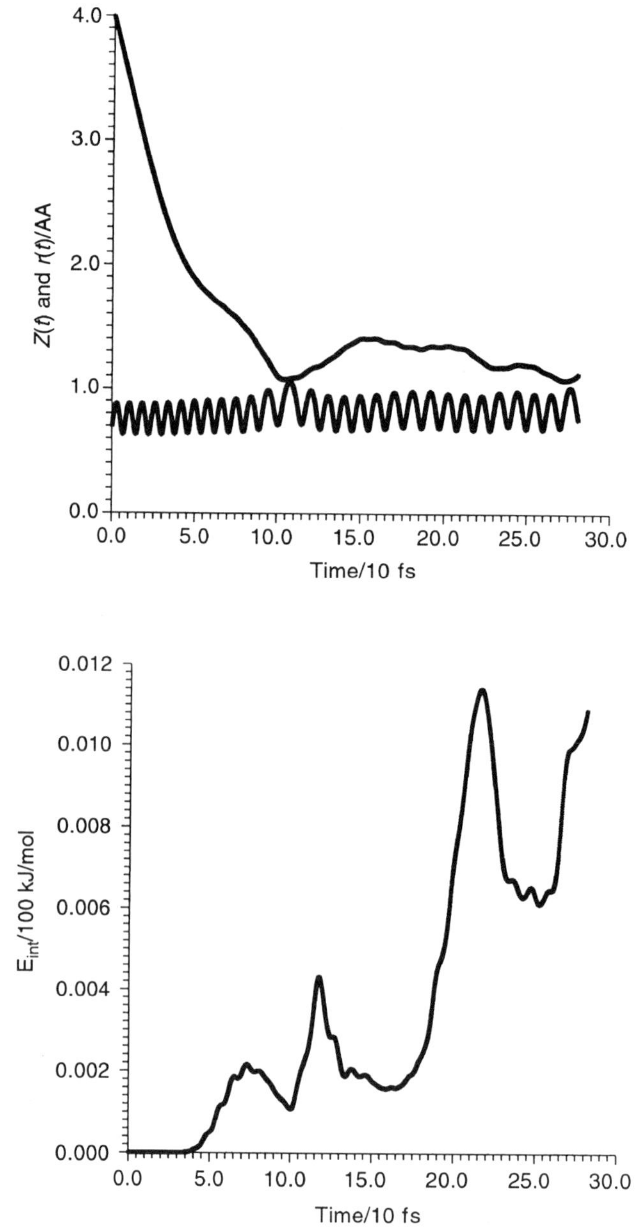

FIGURE 8.6 The bond length of hydrogen $r(t)$ and the center of mass distance $Z(t)$ from the surface as a function of time. The initial kinetic energy is 30 kJ/mol and approach angles $\theta, \phi = 0, 0°$. The molecule is initially in the ground vibrational/rotational state and the surface temperature 800 K. The lower panel shows the energy transfer to the phonons. Distances in Å, energies in units of 100 kJ/mol, and time in 10 fs $= 10^{-14}$ sec.

carried out at $T > 0$K gives rise to a damping factor $\exp(-2W)$, the temperature-dependent Debye-Waller (D-W) factor [65]. The (D-W) factor accounts for the influence of surface vibrations on the diffraction intensities; and for electron and neutron scattering one has used [65]

$$\exp(-\langle \mathbf{u} \cdot \Delta \mathbf{k} \rangle^2) \tag{8.126}$$

where $\mathbf{u}$ is the displacement of the surface atom and $\Delta \mathbf{k}$ the change in wavevector of the scattered particle. The brackets indicate an average over the phonon distribution. This formula leads (see exercise 8.3) to the following approximate expression

$$\exp\left(-\frac{3\hbar^2 \Delta k_z^2 T_s}{m_\alpha k_B \Theta_D^2}\right) \tag{8.127}$$

where T_s is the surface temperature, m_α the mass of the surface atom, Δk_z the change in wavevector in the z-direction, and Θ_D the surface Debye temperature. Introducing the fact that Δk_z, in the limit of large mass difference, is $\Delta k_z \sim 2mv_0 \cos\theta_i/\hbar$ we get the D-W factor

$$2W = \frac{24 m E_0 \cos^2\theta_i T_s}{m_\alpha k_B \Theta_D^2} \tag{8.128}$$

where θ_i is the incident angle and E_0 the initial kinetic energy. This expression can roughly account for the experimentally measured attenuation and broadening of the intensity as a function of surface temperature [37]. We notice that this D-W factor is independent of the potential parameters—in contrast to the factor derived from a dynamical theory below.

By introducing the D-W factor, the experimental data are interpreted approximately such that

$$I_{mn}(T) = \exp(-2W) I_{mn}^0 \tag{8.129}$$

where I_{mn} is the intensity of the (mn) diffraction channel at temperature T and I_{mn}^0 the intensity for a rigid nonvibrating surface (the $T = 0$ K case).

As we have seen above, it is possible to formulate a dynamical approach which in a simple manner includes the surface temperature effect. However, schemes which introduce the Debye-Waller factor as a simple multiplicative factor as illustrated by eq. (8.129) are, of course, even simpler to use.

For surface scattering the derivation of a D-W factor has been considered by a number of researchers (see, e.g., refs. [67, 68, 69]). Previously the theory had been primarily developed for including the effect of inelastic motion in the interpretation of X-ray and neutron scattering data [70].

Thus the probability of interest is that of no energy loss to phonons. In the semiclassical model (considered in section 8.3.1) this probability is given as

$$P_{n \to n} = \frac{\exp(-\rho)}{n!} L_n^0(\rho)^2 \tag{8.130}$$

Introducing the initial distribution of quantum states in a given mode as $P_{n_k} = (1 - z_k)z_k^{n_k}$, where $z_k = \exp(-\hbar\omega_k/k_B T)$, we get the thermal probability as

$$\exp(-2W_k) = \sum_{n_k} P_{n_k \to n_k} z_k^{n_k}(1 - z_k)$$
$$= \exp(-\rho_k(1 + z_k)/(1 - z_k)) I_0(2\rho_k\sqrt{z_k}/(1 - z_k)) \quad (8.131)$$

where $z_k = \hbar\omega_k/k_B T$. For the crystal as a whole we have

$$\exp(-2W) = \Pi_k \exp(-2W_k) \quad (8.132)$$

If the "excitation strength" of phonon mode k is small, i.e., $\rho_k \ll 1$, we can make the approximation $I_0 \sim 1$ and obtain

$$\exp(-2W) \sim \exp\left(-\sum_k \rho_k \coth(\hbar\omega_k/2k_B T_s)\right) \quad (8.133)$$

an expression also obtained in ref. [66]. Introducing the model value for ρ_k (eq. (8.60)) we see that the D-W factor is very sensitive to the steepness of the potential and the initial velocity of the incoming atom. The dependence on the well-depth is, however, more complicated since it is given through the factor $\cot\tau$, i.e., an increase in the well depth may not influence the D-W factor very much. This is in agreement with recent results of Jackson [77] but not with earlier theories [81].

The quantity ρ_k entering the expression for the D-W factor is the excitation strength of mode k. It is related to the mean number of phonon quanta transferred to the mode (see exercise 8.5). The summation over k in eq. (8.133) can be carried out for some model systems (see exercise), but usually it has to be carried out numerically, since ρ_k depends upon the frequencies, and the phonon-distribution function may not be approximated with simple models as the Debye or Einstein model [82]. However, besides this common D-W factor, there are additional temperature dependencies [68], i.e., the diffraction intensities can contain some temperature dependencies which can only be obtained by solving the phonon problem simultaneously with the diffraction problem [75, 76].

EXERCISES

8.3: Discuss the approximations behind the expression (8.127).

8.4: Use the simple collision model for a forced oscillator (see also exercise 8.1) to derive an expression for the Debye-Waller factor. For this model we have

$$\rho = \frac{8\pi^2 m^2 \omega}{\hbar\alpha^2 M}\exp(-2\pi\omega/\alpha v_0) \quad (8.134)$$

where m is the mass of the incoming atom, M the mass of a surface atom, ω the frequency of the "phonon," α the potential steepness parameter, and v_0 the initial velocity of the incoming atom. Assume a Debye frequency distribution

$$n(\omega) = 3\omega^2/\omega_D^3 \tag{8.135}$$

Calculate the D-W factor at surface temperature 200 K and 300 K for a He atom hitting a Cu-surface with kinetic energy 10 kJ/mole. The Debye temperature for Cu is 315 K. Assume that the steepness parameter $\alpha = 3\ \text{Å}^{-1}$. Compare with the value obtained using eq. (8.128). What value of the steepness parameter should have been used in order to obtain the same Debye-Waller factor from the two models?

8.5: Calculate the average number of quanta transferred to the linearly forced oscillator if the oscillator is initially in its ground state.

9

ELECTRON GAS THEORIES

The electronic structure of the surface is obviously important for the interaction and dynamics of molecules with the surface. We have already mentioned, at a qualitative level, the role of the electrons, and more specifically used the electron density when constructing the interaction potentials. The electron density is observable through, for instance, X-ray scattering experiments and scanning tunneling microscopy (STM) images. The electron density plays a central role in the electron gas theories as well as in the density functional theory described in the following chapter.

The simplest treatment of a many-electron system is obtained using the free-electron gas model. For a free-electron gas we calculate the electron density from the number of states in a system with potential equal to zero, i.e., the electron has only kinetic energy. Thus the electron can be described by the simple-particle-in-a-box model.

For the particle in a box with coordinate x between 0 and L_x we have the following normalized solutions

$$\Psi(x) = \sqrt{\frac{2}{L_x}} \sin(k_x x) \tag{9.1}$$

The solution to the particle-in-a-box problem connects the wavenumber k_x to a quantum number through the equation

$$k_x = \frac{\pi}{L_x} n_x \tag{9.2}$$

where n_x takes the values $1, \ldots \infty$ and L_x is the box-length. In three dimensions we then have

$$dk_x dk_y dk_z = \frac{\pi^3}{L_x L_y L_z} dn_x dn_y dn_z \tag{9.3}$$

The energy obtained with the particle-in-a-box eigenfunctions is expressed in terms of the quantum numbers as

$$\varepsilon = \frac{\hbar^2 \pi^2}{2mL^2}(n_x^2 + n_y^2 + n_z^2) \tag{9.4}$$

where we have introduced $L = L_x = L_y = L_z$. The number of states in an energy interval ε to $\varepsilon + d\varepsilon$ can be obtained as $\mathcal{N}(\varepsilon) = d\Omega(\varepsilon)/d\varepsilon$ where $\Omega(\varepsilon)$ is the number of states with energy less than ε. This number is according to eq. (9.4) $\frac{1}{8}$ of the volume of a sphere with radius $\sqrt{2m\varepsilon}L/\pi\hbar$, i.e.,

$$\Omega(\varepsilon) = \frac{V}{6\pi^2}(2m\varepsilon/\hbar^2)^{3/2} \tag{9.5}$$

where $V = L^3$. That is, the density of states $\mathcal{N}(\varepsilon)$ is given as

$$\mathcal{N}(\varepsilon) = \frac{V}{4\pi^2}(2m/\hbar^2)^{3/2}\sqrt{\varepsilon} \tag{9.6}$$

Introducing the density of states per unit volume $n(\varepsilon)$ and the fact that according to the Pauli principle each state can occupy two electrons, we have the density of electrons (number of electrons/unit volume) irrespective of energy given as

$$\overline{n} = 2\int_0^{\varepsilon_F} d\varepsilon n(\varepsilon) = \frac{1}{3\pi^2}(2m/\hbar^2)^{3/2}\varepsilon_F^{3/2} \tag{9.7}$$

where ε_F is the Fermi energy, the highest energy of an electron at 0°K (see Fig. 9.1) and $n(\varepsilon) = \mathcal{N}(\varepsilon)/V$. If the temperature increases one has according to Fermi-Dirac statistics a distribution function

$$f(\varepsilon) = \frac{1}{\exp(\varepsilon - \varepsilon_F)/k_B T + 1} \tag{9.8}$$

where k_B is the Boltzmann constant and ε_F the Fermi energy. The above expression is then modified to

$$\overline{n} = 2\int_0^{\infty} n(\varepsilon) f(\varepsilon) d\varepsilon \tag{9.9}$$

However, at normal temperatures of interest to chemists we have $k_B T \ll \varepsilon_F$. Thus we may approximate the Fermi-Dirac distribution function by the step function, which is unity below and zero above the Fermi level.

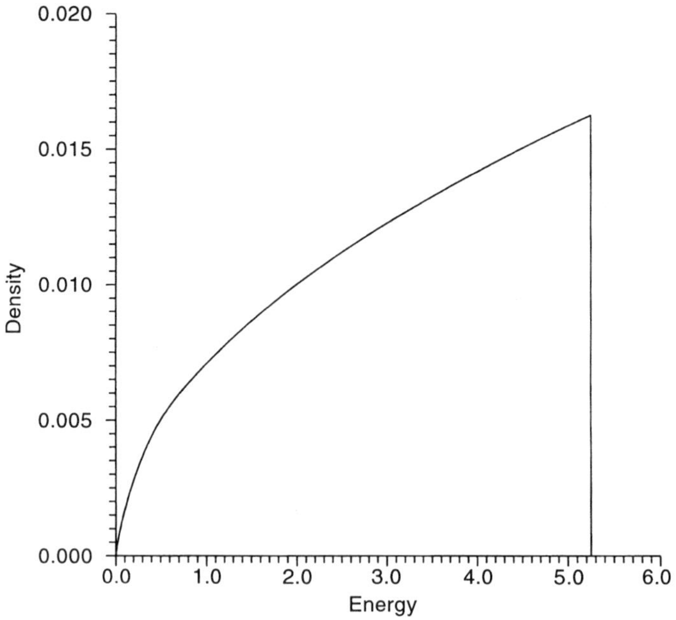

FIGURE 9.1 Density of states per unit volume as a function of energy (in units of 100 kJ/mol) for the free-electron gas. All states up to the Fermi level are occupied.

Alternatively we can obtain the same result by introducing the momenta $p_x = \hbar k_x$, and so on. From eq. (9.3) we have

$$dn_x dn_y dn_z = \frac{V}{\pi^3}\frac{1}{\hbar^3} dp_x\, dp_y\, dp_z \tag{9.10}$$

where $V = L_x L_y L_z$ is the volume available for the electrons. Introducing a maximum momentum p_0 we can express the total number of states as

$$N = \int dn_x dn_y dn_z = \frac{V}{h^3}\int_0^{p_0} p^2 dp \int_0^{\pi} d\theta \sin\theta \int_0^{2\pi} d\phi = \frac{4V\pi}{3h^3} p_0^3 \tag{9.11}$$

where the first integral runs over positive n_x, n_y and n_z values whereas the last runs over positive as well as negative values. It has therefore been multiplied by a factor $\frac{1}{8}$.

Since each state can be occupied by two electrons with opposite spin we get for the ground state the electron density

$$\overline{n} = \frac{2N}{V} = \frac{8\pi}{3h^3} p_0^3 \tag{9.12}$$

The maximum momentum is defined through the Fermi level, i.e.,

$$\varepsilon_F = \frac{p_0^2}{2m} \tag{9.13}$$

where m is the mass of the electron and ε_F the Fermi-energy.

So far, the electrons have been treated as a free-electron gas. In the Hartree and Hartree-Fock theory the electrons are allowed to interact. We shall therefore briefly consider the electron gas in these theories and then introduce the approximate solution in a potential according to the Fermi-Thomas theory [217, 218].

9.1 A FREE-ELECTRON GAS TREATED WITHIN THE HARTREE AND HARTREE-FOCK THEORIES

In order to apply the Hartree and Hartree-Fock expressions given in Appendix C to that of an electron gas we have to introduce a uniform positive charge provided by the ions. That is, the actual charge distribution defined by the core-potentials are simply replaced by a uniform distribution which exactly balances the negative charge from the electron gas. In this model the average charge also is uniform, and hence the effective potential in the Hartree model is zero. Thus in the Hartree theory for the electron gas we have the one-electron Schrödinger equation

$$-\frac{\hbar^2}{2m}\nabla^2\phi_i = \varepsilon_i\phi_i \tag{9.14}$$

with the solution

$$\psi_k(\mathbf{r}) = \frac{1}{\sqrt{L^3}}\exp(i\mathbf{k}\cdot\mathbf{r}) \tag{9.15}$$

with $k_x = 2\pi n_x/L$, and so on, for the running waves. Energy levels are given by

$$\varepsilon(\mathbf{k}) = \frac{\hbar^2}{2m}\frac{4\pi^2}{L^2}(n_x^2 + n_y^2 + n_z^2) \tag{9.16}$$

where n_x, n_y, and n_z now take negative as well as positive values. Comparing with eq. (9.4) we see that (9.16) corresponds to eq. (9.4) with integer quantum numbers, that is, to $\frac{1}{8}$ of the allowed quantum numbers in eq. (9.16). However, since the energies are doubly degenerate in each quantum number we get 8 times as many, i.e., the density of states will be unchanged.

Thus within the Hartree theory we get the same energies as in the zero potential case. The Hartree-Fock case can now be treated by adding the exchange term to the kinetic energy, i.e., we have

$$-\frac{\hbar^2}{2m}\nabla_i^2\phi_i(\mathbf{r}_i) - e^2\sum_j\left[\int\frac{\phi_j^*(\mathbf{r}_j)\phi_i(\mathbf{r}_j)\phi_j(\mathbf{r}_i)}{r_{ij}\phi_i(\mathbf{r}_i)}d\mathbf{r}_j\right]\phi_i(\mathbf{r}_i) = \varepsilon_i\phi_i(\mathbf{r}_i) \tag{9.17}$$

Introducing again the plane wavefunction as "trial" functions we see that we have to evaluate an integral of the type

$$-e^2\sum_j\left[\int\frac{\exp(i(\mathbf{k}_i-\mathbf{k}_j)\cdot(\mathbf{r}_j-\mathbf{r}_i))}{r_{ij}}d\mathbf{r}_j\right] \tag{9.18}$$

which is (see exercise 9.2)

$$\frac{4\pi e^2}{|\mathbf{k}_i-\mathbf{k}_j|^2} \tag{9.19}$$

and hence the exchange energy is given as

$$\varepsilon_x(\mathbf{k}) = -\frac{4\pi e^2}{8\pi^3}\int\frac{d\mathbf{k}'}{|\mathbf{k}-\mathbf{k}'|^2} \tag{9.20}$$

where we have replaced the sum over $\mathbf{k}_j$ with an integral $\int d\mathbf{k}/8\pi^3$. Introducing the maximum value of k as k_0 (here also denoted k_F) we can finally evaluate the integral over $\mathbf{k}$, and we obtain

$$\varepsilon_x(k) = -\frac{e^2k_0}{2\pi}\left[2+\frac{k_0^2-k^2}{kk_0}\ln\left|\frac{k_0+k}{k_0-k}\right|\right] \tag{9.21}$$

and in order to obtain the total energy we have to add the kinetic energy, i.e.,

$$\varepsilon_k = \frac{\hbar^2k^2}{2m} + \varepsilon_x(k) \tag{9.22}$$

where the last term is the correction to the Hartree theory. Considering the band width we note that $\varepsilon(k_0) = \frac{\hbar^2k_0^2}{2m} - \frac{e^2k_0}{\pi}$ and $\varepsilon(0) = -\frac{2e^2k_0}{\pi}$ and hence the band width is obtained as

$$\varepsilon(k_0) - \varepsilon(0) = \frac{\hbar^2k_0^2}{2m} + \frac{e^2k_0}{\pi} \tag{9.23}$$

In the Hartree-Fock theory the energy bands are not only shifted to lower values of the energy, but the band width is increased with the last term of eq. (9.23). Further improvement of the quantum treatment of the electron gas involves inclusion of correlation energy. This is, however, somewhat more difficult and outside the scope of this presentation, but we may mention that for very low densities the average correlation energy has been estimated to $-9e^2/10r_s$ where r_s is defined through the connection to the Fermi energy

$$\varepsilon_F = \frac{\hbar^2}{2m}(9\pi/4)^{2/3}\frac{1}{r_s^2} \tag{9.24}$$

Early attempts to evaluate the correlation energy in the high-density limit by using perturbation theory are due to Wigner [216].

Another simple model for the electrons in metals is the so-called jellium model, in which the wavefunction is an eigenfunction to the free electron wavefunction below the surface and has an exponential decay in the gas phase. Thus one assumes the hamiltonian given as

$$H_{el} = -\frac{\hbar^2}{2m}\nabla^2 \text{ for } z > 0 \text{ and}$$

$$H_{el} = -\frac{\hbar^2}{2m}\nabla^2 - V_s \text{ for } z < 0 \tag{9.25}$$

where the surface or, strictly speaking, the jellium edge is located at $z = 0$ and V_s is a constant average potential for the conduction electrons. A solution to the Schrödinger equation which has continuous first derivative at $z = 0$ is

$$\psi_{k_x k_y k_z} = c_k \exp(i(k_x x + k_y y))\exp(-K_s z) \text{ for } z > 0 \tag{9.26}$$

$$\psi_{k_x k_y k_z} = c_k \exp(i(k_x x + k_y y))\left(\cos(k_z z) - \frac{K_s}{k_z}\sin(k_z z)\right) \text{ for } z < 0 \tag{9.27}$$

where $K_s = \sqrt{2mV_s/\hbar^2 - k_z^2}$, c_k is a normalization factor, and $E = (\hbar^2/2m)(k_x^2 + k_y^2 + k_z^2)$. The potential V_s is a parameter which can be fitted to properties of the bulk metal, for instance, the average electron density. Often the average electron density ρ is related to the volume of a sphere with radius r_s such that

$$\frac{4}{3}\pi r_s^3 = \rho^{-1} \tag{9.28}$$

where ρ is given, in say, electrons/Å^3 (see Table 9.1). Table 9.1 shows that the electron density, if plotted against the number of valence electrons, has a maximum for metals with about half-filled d-levels, i.e., 6–8 valence electrons. This is in agreement with the reactivity (ability to enhance surface reactions) expected from the simple concepts mentioned in the introduction. Figure 9.2 shows the density of states $\rho(z)$ obtained with the jellium model

$$\rho(z) = \int_0^{\varepsilon_F} dE \sum_{n_x n_y n_z} |\psi(x, y, z)|^2 \delta(E - E_{n_x n_y n_z}) \tag{9.29}$$

TABLE 9.1 Electron Densities for Some Metals in Units of Electrons/Å^3

Metal	ρ	n_v	Metal	ρ	n_v
Na	0.033	1	Cr	0.338	6
K	0.020	1	Mo	0.310	6
Rb	0.016	2	Mn	0.347	7
Ca	0.059	2	Tc	0.315	7
Sr	0.047	3	Ru	0.316	8
Sc	0.121	3	Fe	0.354	8
Y	0.106	3	Co	0.336	9
Zr	0.174	4	Cu	0.243	11
Ti	0.198	4	Ni	0.343	10
V	0.281	5	Pd	0.228	10
Nb	0.243	5	Ag	0.169	11

n_v is the number of valence electrons. Data from ref. [219].

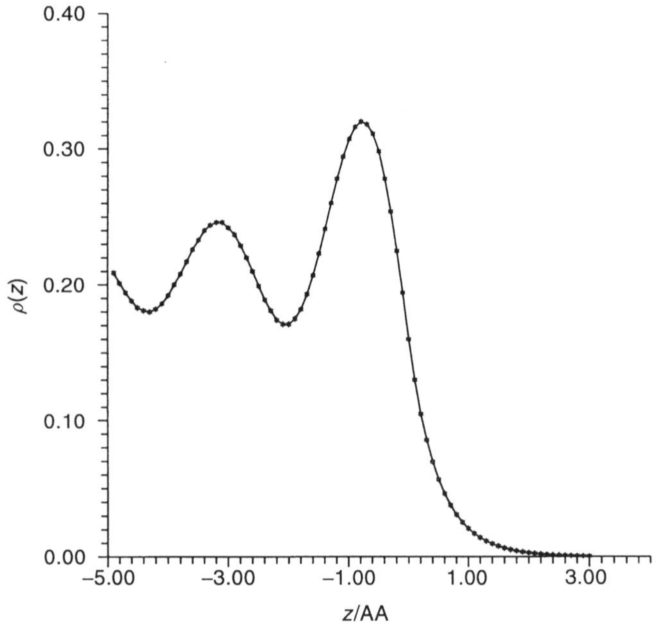

FIGURE 9.2 Density of states $\rho(z)$ as a function of z near the surface edge ($z = 0$) for the jellium model. We notice that charge density spills over into the vacuum up to about 2–3 Å.

We notice oscillations near the surface edge and the exponential tail in the vaccum. If an atom or a molecule is adsorbed, the density of states near the surface will change. To a first approximation, this problem can be dealt with by adding an extra potential term to the jellium hamiltonian [182].

Instead of using $V = 0$ as in the simple electron gas theory, or a constant, as in the jellium theory, a pseudo-potential $V(\mathbf{R}, z)$ can be used to calculate the electronic wavefunction

$$\left[-\frac{\hbar^2}{2m}\nabla^2 + V(\mathbf{R}, z)\right] \Psi(\mathbf{R}, z) = E\Psi(\mathbf{R}, z) \tag{9.30}$$

Due to the periodicity of the potential, i.e., $V(\mathbf{R}, z) = V(\mathbf{R} + \mathbf{R}_l, z)$ where $\mathbf{R}_l$ is a surface lattice vector, also the wavefunction is periodic. Such solutions are called Bloch wave functions and will be used in Chapter 11 when discussing electron-hole pair excitation. More generally the solution to eq. (9.30) depends on a potential which is a function of the electron density itself — that is, a Hartree potential (see eq. (9.17)) added to an exchange-correlation potential. The systematic formulation of this is obtained through the density functional theory (DFT), to be considered in the next chapter. Simpler corrections to the jellium model are devised through the effective medium approach [210]. In the effective medium approach the exact interaction between the adsorbate and the substrate is replaced by the interaction to a jellium surface, i.e., the adsorbate is embedded in an electron gas. This scheme gives the energy in terms of the embedding energy, which is a function of the electron density. Various corrections for core-core repulsion and d-band corrections have then been added in order to improve the description of the molecule-surface interaction obtained with the effective medium approach [244].

9.2 THE FERMI-THOMAS THEORY

The concept of screening, i.e., the ability of an electron gas to screen the positive ionic charge or background, plays a role in simplistic treatments of atom/molecule-surface interaction. The simplest way of treating this at a semi-quantitative level is through the Fermi-Thomas theory.

If the electron moves in a potential $V(\mathbf{r})$, the total energy is a sum of the kinetic and potential energy

$$E = \frac{p_0^2}{2m} - eV(\mathbf{r}) \tag{9.31}$$

i.e., in this case, using eq. (9.12) the electron density would be

$$\overline{n}(\mathbf{r}) = \frac{8\pi}{3h^3}(2m(E + eV(\mathbf{r})))^{3/2} \tag{9.32}$$

A fundamental equation in electrostatic theory is the Poisson equation. The Poisson equation can be derived from the Gauss theorem [245] and relates the potential to the charge density or distribution by the equation

$$\nabla^2 V(\mathbf{r}) = -4\pi\rho(\mathbf{r}) \tag{9.33}$$

Numerical techniques for solving the Poisson equation for an arbitrary charge distribution exist (see, e.g., [246]). Here we shall just consider the simple case where the distribution is given by the expression (9.32). Using the fact that $\rho(\mathbf{r}) = -e\overline{n}(\mathbf{r})$ we can obtain the Fermi-Thomas equation [248]

$$\nabla^2 V(\mathbf{r}) = 4\pi e\overline{n}(\mathbf{r}) = \frac{32e\pi^2}{3h^3}(2m(E + eV(\mathbf{r})))^{3/2} \tag{9.34}$$

Introducing now the electron density for a Fermi gas (from eqs. (9.12) and (9.13)) we have

$$n_0 = \frac{8\pi}{3h^3}(2m\varepsilon_F)^{3/2} \tag{9.35}$$

and considering electrons with total energy near the Fermi limit we have

$$\nabla^2 V = 4\pi e n_0 \left(1 + \frac{eV}{\varepsilon_F}\right)^{3/2} \tag{9.36}$$

Consider now an electron gas with a uniform positive background from ions with a charge density n_0 to balance the negative electronic density. The Poisson equation for this system is

$$\nabla^2 V(\mathbf{r}) = 4\pi e(\overline{n}(\mathbf{r}) - n_0) \tag{9.37}$$

or using that from eqs. (9.32) and (9.35) we have $\overline{n} = n_0(1 + eV/\varepsilon_F)^{3/2}$. If we now linearize this expression, i.e., use $(1 + eV/\varepsilon_F)^{3/2} \sim 1 + \frac{3}{2}eV/\varepsilon_F$ and assume a spherical symmetric potential, we get

$$\nabla^2 V = \frac{d^2V(r)}{dr^2} + \frac{2}{r}\frac{dV(r)}{dr} = \frac{6\pi e^2 V(r)}{\varepsilon_F} n_0 \tag{9.38}$$

with the solution

$$V(r) = \frac{q}{r}\exp(-ar) \tag{9.39}$$

i.e., a screened Coulomb potential, the screening factor being $\exp(-ar)$ with a screening constant

$$a = \frac{4\pi e}{h}\sqrt{\frac{m}{\pi}k_F} = 2\sqrt{\frac{k_F}{\pi a_0}} \tag{9.40}$$

where $a_0 = \hbar^2/me^2$ is the Bohr radius and k_F the Fermi wavevector. The expression was obtained using eq. (9.35) and $\varepsilon_F = \hbar^2 k_F^2/2m$.

It is now convenient to introduce the radius r_0, defined through

$$\frac{1}{n_0} = \frac{4}{3}\pi r_0^3 \tag{9.41}$$

If we use the fact that $k_F = (3\pi^2 n_0)^{1/3}$, this gives the following expression for the screening length $1/a$

$$1/a = \tfrac{1}{2}(2\pi/3)^{1/3}\sqrt{a_0 r_0} \tag{9.42}$$

Thus according to the Fermi-Thomas theory a charge q embedded in an electron gas with a positive uniform background feels a screened Coulomb potential. Typical r_0 values for metals are between 1 Å and 3 Å, giving values of $1/a$ between 0.5 and 1 Å. However, the more accurate theory based on a plasma model for the electron gas gives a screening length of about twice (see exercise 9.4) that obtained by the Fermi-Thomas theory [248]. In the plasma model for the electron gas one obtains the following expression for the screened potential

$$\frac{e^2}{r}\left(1 - \frac{2}{\pi} Si(k_c r)\right) \tag{9.43}$$

where $k_c = \omega_p / v_0$ and the function $Si(x)$ is defined by

$$Si(x) = \int_0^x dy \frac{\sin(y)}{y} \tag{9.44}$$

ω_p is the frequency for plasma oscillations $\omega_p = \sqrt{4\pi e^2 n/m}$, n the electron density, and v_0 the velocity of electrons near the Fermi level.

Returning to the Thomas-Fermi potential, it satisfies, as we have seen, the classical Poisson equation

$$\nabla^2 V = a^2 V \tag{9.45}$$

where $a^2 = (4k_F/\pi a_0)$. Considering the Fourier transform of the Thomas-Fermi potential, i.e.,

$$\hat{V}(\mathbf{k}) = \int d\mathbf{r} V(\mathbf{r}) \exp(i\mathbf{k}\cdot\mathbf{r}) \tag{9.46}$$

we obtain the Fourier coefficients

$$\hat{V}_{TF}(k) = \frac{4\pi q}{k^2 + a^2} = \frac{4\pi q}{k^2 \varepsilon(k)} \tag{9.47}$$

where the factor $4\pi q/k^2$ would be obtained with just the Coulomb potential, i.e., the correction $\varepsilon(k) = 1 + a^2/k^2$ plays the role of a dielectric constant.

Refinements of the Thomas-Fermi theory have been considered, for example, by March and Murray [201]. By treating scattering of electron waves from a test charge using first-order wave theory, they obtained the following modified Poisson equation

$$\nabla^2 V = \frac{2me^2 k_F^2}{\hbar^2 \pi^2} \int d\mathbf{r}' \frac{j_1(2k_F|\mathbf{r} - \mathbf{r}'|)}{|\mathbf{r} - \mathbf{r}'|^2} V(\mathbf{r}') \tag{9.48}$$

where $j_1(x) = (\sin x - x\cos x)/x^2$. Considering the Fourier transform of the solution to this equation one obtains [201]

$$\hat{V}_{MM} = \frac{4\pi q}{k^2 + \frac{k_F g(k/2k_F)}{\pi a_0}} \tag{9.49}$$

where the function $g(x)$ is defined by

$$g(x) = 2 + \frac{x^2 - 1}{x} \ln \left| \frac{1-x}{1+x} \right| \tag{9.50}$$

with $x = k/2k_F$. Since for $x \to 0$ we have $g(x) \to 4$ we recover the Thomas-Fermi (TF) result in that limit. The dielectric function obtained from this theory is [202]

$$\varepsilon(k) = 1 + \frac{k_F}{\pi a_0 k^2} g(k/2k_F) \tag{9.51}$$

These expressions show that the TF theory is limited in validity to the long wavelength region, where $k/2k_F \ll 1$.

The classical Thomas-Fermi theory is a statistical theory which allows the electrons to move independently of each other. The one-electron wavefunctions are obtained using the classical electrostatic potential only. Corrections to the simple theory are obtained by introducing an exchange and a correlation correction. Also a gradient correction has been introduced to correct for the use of a free-electron gas kinetic energy in the original theory [203, 204].

The energy functional, i.e., the energy expressed in terms of the density $n(\mathbf{r})$ in the TFD (Thomas-Fermi-Dirac) model is expressed as [206]

$$E = c_k \int n^{5/3} d\mathbf{r} + \int n V_N d\mathbf{r} + \frac{1}{2} e^2 \int \frac{n(\mathbf{r}) n(\mathbf{r}')}{|\mathbf{r} - \mathbf{r}'|} - c_s \int n^{4/3} d\mathbf{r} \tag{9.52}$$

where the first term with $c_k = (3\hbar^2/10m)(3\pi^2)^{2/3}$ is the kinetic energy functional, the second term contains the interaction between the charge distribution and the external potential from the nuclei, the third term is the classical electron-electron interaction, and the final term is the Dirac exchange term with $c_s = \frac{3e^2}{4}(3/\pi)^{1/3}$. For finite systems, a correction factor for "self-exchange" has been introduced by Rae [205]. Molecular binding cannot be described by a local form of the kinetic energy—but requires some delocalization. Although gradient corrections to the local form can be introduced in an attempt to improve the situation, the theory has in fact been replaced by the more accurate density functional theory.

Suppose that we now require that the energy should be minimized subject to the constraint that the integration of the density over space should give the total number of electrons in the system

$$\int d\mathbf{r} n(\mathbf{r}) = N \tag{9.53}$$

We have the following variational principle

$$\delta(E - \mu N) = 0 \tag{9.54}$$

giving an expression for the Lagrange multiplier μ (the chemical potential) as

$$\mu = \frac{5}{3} c_k [n(\mathbf{r}]^{2/3} + V_N(\mathbf{r}) + e^2 \int \frac{n(\mathbf{r}')}{|\mathbf{r} - \mathbf{r}'|} d\mathbf{r}' - \frac{4}{3} c_s [n(\mathbf{r}]^{1/3} \tag{9.55}$$

If we drop the last term (the exchange term) and combine the nuclear and electron potential to $V(\mathbf{r})$ we get

$$n(\mathbf{r}) = \frac{8\pi}{3h^3}(2m(\mu - V(\mathbf{r}))^{3/2} \tag{9.56}$$

i.e., the expression (9.32).

The electron density is defined in terms of effective one-electron wavefunctions as

$$n(\mathbf{r}) = 2 \sum_i^{N/2} |\psi_i|^2 \tag{9.57}$$

where the sum runs over the occupied orbitals. In the Thomas-Fermi theory no explicit use is made of the wavefunctions, and corrections are introduced as gradient corrections to the density. The concept of electron density plays a role in many approximate theories of atom/molecule-surface interaction. We have previously mentioned the EAM (embedded atom), the EDIM (embedded diatom in molecules), and the effective medium (EM) theories. Electron gas theories have been used for the calculation of atomic and molecular properties [201], for an estimate of the short-range potential between molecules [207], and also for atom-surface interaction potentials [208]. But even more important is the fact that an in-principle exact formulation of the electron structure problem for many electron problems can also be expressed in terms of the density. This formulation will briefly be reviewed in the next chapter.

EXERCISES

9.1: Calculate the Fermi energy for copper with molar mass 63.5, one conduction electron contributing to the electron gas, and a density of 8.96 g/cm^3. *Hint*: Use eq. (9.35) and that the electron mass is 0.0005486 amu.

9.2: Derive the eqs. (9.19) and (9.21). Use the fact that a charge distribution $f(\mathbf{r})$ gives rise to a potential which satisfies the Poisson equation

$$\nabla^2\phi(\mathbf{r}) = -4\pi f(\mathbf{r}) \tag{9.58}$$

and that

$$\int x \ln(a \pm x) dx = \frac{1}{2}(x^2 - a^2)\ln(a \pm x) - \frac{1}{2}\left(\frac{1}{2}x^2 \mp ax\right) \tag{9.59}$$

9.3: Show that the kinetic energy density functional for a free-electron gas is

$$E_k = \frac{3\hbar^2}{10m}(3\pi)^{2/3}\int n(\mathbf{r})^{5/3}d\mathbf{r} \tag{9.60}$$

Hint: Use the particle-in-a-box wavefunctions.

9.4: Compare the screening length of the plasma model with that of the Thomas-Fermi model, i.e., $1/a$ with $1/k_c$.

10

DENSITY FUNCTIONAL THEORY

The density functional theory (DFT) was mentioned previously in Chapter 4, and the Thomas-Fermi method can also be viewed as a special case of DFT. DFT has become very popular over the last 20 years or so. The reasons are obvious — it scales as N^3 where N is the number of electrons — by contrast to standard ab initio Hartree-Fock theory, which scales as N^4, and the results are more accurate (not to mention accurate ab initio theories, which scale as N^7). Furthermore the DFT theory operates in three dimensions (x, y, z) in which the electron density is defined — no matter how many electrons are involved. DFT theory has been reviewed and described in numerous books and review articles; see, for instance, [212]. We shall therefore only give a brief presentation of it here.

For a given external field from the nuclei $v(\mathbf{r})$ the energy is according to the fundamental theorems by Hohenberg and Kohn [211] a functional of the density, i.e., the ground state energy is given as

$$E(n(\mathbf{r})) = F(n(\mathbf{r})) + \int v(\mathbf{r})n(\mathbf{r})d\mathbf{r} \tag{10.1}$$

where the contribution to $F(n)$ can be partitioned in a contribution from the kinetic energy, exchange-correlation, and classical electron-electron interaction. Since the exchange-correlation (xc) term is not known, the price which is paid for the simplification is that there does not exist a systematic (first principle) way of improving a given result. But it is believed that a certain transferability of the (xc) term exists. That is, once a suitable term is found for a certain class of problems the same term will predict properties of the remaining problems of that class. The DFT method replaces the many-electron problem with a set of one-electron equations.

In DFT one works, as the name indicates, with the density of the electrons rather than the wavefunction for the electrons themselves. Thus the energy of the ground state is expressed as a functional $E(n(\mathbf{r}))$ of the electron density $n(\mathbf{r})$. The variational principle with respect to the density gives a set of self-consistent equations, which, however, have to be solved iteratively (see below). The energy functional is, as mentioned, split into several contributions

$$E(n(\mathbf{r})) = E_{kin} + \int v(\mathbf{r})n(\mathbf{r})d\mathbf{r} + \frac{e^2}{2}\int \frac{n(\mathbf{r})n(\mathbf{r}')}{|\mathbf{r}-\mathbf{r}'|}d\mathbf{r}d\mathbf{r}' + G_{xc}(n(\mathbf{r})) \tag{10.2}$$

where the first term is the kinetic energy contribution, the second involves the external potential including electron-ion

$$v(\mathbf{r}) = -\sum_{\alpha} \frac{e^2 Z_\alpha}{|\mathbf{R}_\alpha - \mathbf{r}|} \tag{10.3}$$

and eventually also ion-ion interaction, the third term is the classical electron-electron Coulomb repulsion, and the last the contribution from "nonclassical" effects correlation and exchange. The computation of Coulomb energies entering the second term is difficult due to the long range of the interaction. This problem is solved by transforming to the reciprocal space and using a method developed by Ewald [213] (see also section 6.6).

In terms of doubly occupied electronic states ψ_i we now have

$$E_{kin} = -2\frac{\hbar^2}{2m}\sum_{i=1}^{N/2}\int \psi_i \nabla^2 \psi_i d\mathbf{r} \tag{10.4}$$

and the electronic density is

$$n(\mathbf{r}) = 2\sum_{i}^{N/2} |\psi_i(\mathbf{r})|^2 \tag{10.5}$$

where N is the number of electrons and i runs over the number of occupied orbitals. Below we shall divide the density n in two spin components n_+ and n_-.

The so-called Kohn-Sham self-consistent equations [215] are obtained by minimizing the energy functional with respect to the orbital ψ_i, subject to the constraint that the density is fixed. Thus one gets [215]

$$\left[-\frac{\hbar^2}{2m}\nabla^2 + v(\mathbf{r}) + e^2\int \frac{n(\mathbf{r}')}{|\mathbf{r}-\mathbf{r}'|}d\mathbf{r}' + v_{xc}\right]\psi_i(\mathbf{r}) = \varepsilon_i \psi_i(\mathbf{r}) \tag{10.6}$$

where the potential v_{xc} formally is given by

$$v_{xc}(\mathbf{r}) = \frac{\delta G_{xc}(n(\mathbf{r}))}{\delta n(\mathbf{r})} \tag{10.7}$$

Thus this version of the theory works with one-electron wavefunctions determined by an effective potential

$$v_{eff}(\mathbf{r}) = v_{el}(\mathbf{r}) + v_{xc}(\mathbf{r}) \tag{10.8}$$

where the first term is the total electrostatic potential and the last the exchange-correlation contribution, which depends in a complicated manner on the density. Complete knowledge of the term would require a solution of the many-body electron problem. Since this is what one tries to avoid in the DFT approach, the trick is to find a good parametrization of this term. Numerous suggestions exist in the literature, and the acquisition of practical experience with the various parametrizations is an active research area at present. An obvious first estimate of the G_{xc} contribution is to use the results of a uniform electron gas, such that

$$G_{xc} \sim G_{uni} = \int d\mathbf{r} g_0(\mathbf{r}) \tag{10.9}$$

where approximations for $g_0(\mathbf{r})$ are obtained by, for example, Wigner [216], Gell-Mann and Brueckner [220], Nozieres and Pines [221], and many others.

Since the density depends on ψ_i the Kohn-Sham equations are solved iteratively. Thus an initial guess of the density is needed. In practice any reasonable estimate will do, for instance, an estimate obtained from the tabulated values of Clementi et al. [142].

In the above equations, only the last term, the "exchange-correlation," term, poses problems and prevents straightforward applications. If this term were known, then the ground state energy would be obtainable as well as the spin densities n_+ and n_- using simple basis set expansion and matrix diagonalization techniques. Thus the accuracy which can be obtained is limited by the parametrization of this term.

However, the important fact is that the term is still a functional of the density and therefore approximate expressions can be introduced for $G_{xc}(n)$. A number of approximate forms have been suggested. The simplest is to introduce the free-electron gas expression for this term. In the local density approximation (LDA), one assumes, as mentioned, that the exchange-correlation energy locally is given as that of a homogeneous electron gas

$$G_{xc}(n_+, n_-) = \int d\mathbf{r} n(\mathbf{r}) \varepsilon_{xc}^{unif}(n_+, n_-) \tag{10.10}$$

where the density has been divided in its two spin components n_+ and n_- and the exchange-correlation energy per electron of a uniform electron gas is ε_{xc}^{unif}.

An LDA parametrization of the exchange-correlation energy has been given by several groups [222, 223, 224]. The LDA approximation has been used successfully in solid state physics to obtain structure, dispersion curves, phonon-spectra, and so on. Although the LDA method has been used successfully for such systems, it usually does a poor job in predicting the barrier heights for chemical reactions at surfaces. The LDA underestimates the barriers for dissociation, but overestimates chemisorption energies. Typical barriers are in the 0.1 – 1 eV range and are needed with an accuracy of better than about 0.01–0.02 eV to be of practical value for use in dynamics calculations. Since the LDA method cannot be used to estimate the correlation energy (see, e.g., [226]), the reaction barriers are typically not well estimated in the LDA approximation. Better estimates are available and obtained through the generalized gradient approximations (GGA) [227, 89, 230]. Thus the accuracy of present functionals is believed to be about 0.1–0.2 eV, i.e., a bit above the requirements for chemical reactions. However, more experience with this particular application might easily bring the error down to a level which would be acceptable from a practical point of view.

10.1 EXCHANGE-CORRELATION PARAMETRIZATION

A number of different parametrizations of the exchange-correlation term exist. Here we mention a relatively simple scheme by Perdew et al. Thus the GGA method has been parametrized as

$$G_{xc}(n_+, n_-) = \int d\mathbf{r} n(\mathbf{r}) F(n_+, n_-, \nabla n_+, \nabla n_-) \tag{10.11}$$

The simple parametrization given by Perdew et al. in ref. [228] for the correlation part in the GGA scheme is expressed as

$$G_c(n_+, n_-) = \int d\mathbf{r} n(\mathbf{r}) F_c(r_s, \eta, t) \tag{10.12}$$

where $\eta = (n_+ - n_-)/n$ is the spin polarization, $n = n_+ + n_-$ and

$$F_c(r_s, \eta, t) = \varepsilon_c^{unif}(r_s, \eta) + H(r_s, \eta, t) \tag{10.13}$$

The first term is the correlation energy of the uniform electron gas [225] expressed in terms of η and $r_s = (3/(4\pi n))^{1/3}$ the Seitz radius. In ref. [225], a simple parametrization of the correlation energy of the first term for an uniform electron gas has been given.

The parameter t is expressed in terms of the Thomas-Fermi screening wave number $a = \sqrt{4k_F/\pi a_0}$ (see also Chapter 9), the spin scaling factor

$$\phi(\eta) = ((1+\eta)^{2/3} + (1-\eta)^{2/3})/2 \tag{10.14}$$

and

$$t = \frac{|\nabla n|}{2\phi a n} \tag{10.15}$$

The function H should interpolate smoothly between the low gradient or high density limit $t \to 0$ and the opposite limit where the correlation should vanish, this opposite limit being dominated by the exchange term. Both limits could be obtained by the ansatz

$$H = \frac{e^2}{a_0}\gamma\phi^3 \ln\left[1 + \frac{\beta}{\gamma}t^2(1 + At^2)/(1 + At^2 + A^2t^4)\right] \tag{10.16}$$

where

$$A = \frac{\beta}{\gamma}/(\exp(-\varepsilon_c^{unif}/(\gamma\phi^3 e^2/a_0)) - 1) \tag{10.17}$$

$\beta = 0.066725$, and $\gamma = 0.031091$ (see ref. [228]).

The exchange contribution has also been parametrized in a simple fashion

$$G_x = \int d\mathbf{r} n(\mathbf{r})\varepsilon_x^{unif}(n)F_x(s) \tag{10.18}$$

where the exchange energy of a uniform gas is given as

$$\varepsilon_x^{unif} = -\frac{3e^2 k_F}{4\pi} \tag{10.19}$$

and the correction function

$$F_x(s) = 1 + \kappa - \kappa/(1 + \mu s^2/\kappa) \tag{10.20}$$

where $\kappa = 0.804$, $\mu = 0.21951$, and the parameter s is defined as

$$s = \frac{|\nabla n|}{2k_F n} \tag{10.21}$$

where k_F is the Fermi wavevector. The parameter range of interest is $0 \le s \le 3$ and $0 \le r_s/a_0 \le 10$, for which the parametrization has been shown to work as well as others for determining the binding energy of a series of diatomic and polyatomic molecules [228].

As can be imagined from the above discussion, the exchange-correlation contribution is not known a priori and it is not clear whether a systematic improvement of the results by changing the GGA is generally achievable [232]. A pragmatic approach is to incorporate more parameters in this term and then to determine those by fitting to a series of known experimental or high-level ab initio data [233].

The brief presentation given here demonstrates that further corrections are certainly possible and that some caution has to be expressed as to the accuracy which can be obtained using density functional theory, at least at present.

10.2 REPRESENTATIONS

Neglect of both the exchange and correlation terms will correspond to a Hartree approximation in traditional ab initio language. If only the correlation term is neglected then the approach corresponds to the Hartree-Fock theory. It has been suggested [231] to replace the exchange term by a local approximation

$$E_x(n) = -\frac{18}{24\pi}(3\pi^2)^{1/3} \int d\mathbf{r} n^{4/3}(\mathbf{r}) \tag{10.22}$$

which gives the following simple approximate equation for the wavefunction $\psi_i(\mathbf{r})$

$$\left[-\frac{\hbar^2}{2m}\nabla^2 + e^2 \int \frac{n(\mathbf{r}')}{|\mathbf{r} - \mathbf{r}'|} d\mathbf{r}' - \frac{e^2}{\pi}(3\pi^2 n(\mathbf{r}))^{1/3} \right] \psi_i(\mathbf{r}) = \varepsilon_i \psi_i(\mathbf{r}) \tag{10.23}$$

These equations are the density functional version of the ab initio Slater $X\alpha$ method (with a multiplicative parameter α on the exchange term). The exchange potential is similar to that originally given by Gaspar [234].

To facilitate the treatment of solids and molecule-surface interaction, it is necessary to use the density functional treatment only for the metal valence electrons. The electron-ion core interaction can be incorporated using pseudopotentials [235]. Hereby the strong electron ion interaction is replaced by the much weaker valence electron effective core potential, sometimes including relativistic effects [238].

The Kohn-Sham equations can be solved by expanding the wavefunction ψ_i in a suitable basis set, such as, for example, the plane wave basis

$$\psi_i(\mathbf{r}) = \exp(i\mathbf{k} \cdot \mathbf{r}) \sum_G c_{iG} \exp(i\mathbf{G} \cdot \mathbf{r}) \tag{10.24}$$

where $\mathbf{G}$ are reciprocal lattice vectors. Inserting this expression in the above equation (10.5) we obtain from eq. (10.2) a matrix equation in the coefficients c_{iG}. In the plane wave basis set the kinetic energy is diagonal. Thus we have

$$\sum_G \left[\frac{\hbar^2}{2m}(k+G)^2 \delta_{G,G'} - \sum_\alpha e^2 Z_\alpha \int d\mathbf{r} \frac{\exp(i(\mathbf{G}-\mathbf{G}')\cdot\mathbf{r})\, e^2}{|\mathbf{R}_\alpha - \mathbf{r}|} \frac{e^2}{2} \int d\mathbf{r} \int d\mathbf{r}' \frac{n(\mathbf{r}')\exp(i(\mathbf{G}-\mathbf{G}')\cdot\mathbf{r})}{|\mathbf{r}-\mathbf{r}'|} + v_{xc}(\mathbf{G}-\mathbf{G}') \right] c_{i,G} = \varepsilon_i c_{i,G'} \tag{10.25}$$

However, the wavefunction may instead be expanded on a grid and the kinetic energy evaluated using the FFT (fast Fourier transform) method. In the grid representation the potential matrices are "diagonal," i.e., the operation of the potential on a wavefunction is a simple multiplication with the value at the grid point.

In any case, both the plane wave and the grid methods result in large matrix diagonalization. Since only a fraction of the eigenvalues are needed, various efficient techniques for the truncation of the matrices and the problem can be introduced. The DFT theory gives the lowest adiabatic potential energy surface and it has, as mentioned in Chapter 4, been extensively used in calculations on molecule-surface interaction potentials. The DFT methodology has therefore given us an important step forward for the understanding of the influence of structure and other surface characteristics on surface reactions. Density functional methods are, therefore, a part of standard program packages [236].

10.2.1 The Fast Fourier Transform Method

In order to evaluate the effect of the kinetic energy operator operating on the wavefunction it is convenient to use the fast Fourier transform (FFT) methodology [237]. The kinetic energy operator entering the DFT problem is

$$\nabla^2 = \frac{\partial^2}{\partial x^2} + \frac{\partial^2}{\partial y^2} + \frac{\partial^2}{\partial z^2} \tag{10.26}$$

Using a discrete representation of the wavefunction $\Psi(x_i, y_j, z_k)$ at $N = N_x N_y N_z$ grid points we can introduce the Fourier transform as

$$\Psi(x_i, y_j, z_k) = \frac{1}{\sqrt{N}} \sum_{l=-N_x/2}^{N_x/2-1} \sum_{m=-N_y/2}^{N_y/2-1} \sum_{n=-N_z/2}^{N_z/2-1} c_{lmn} \exp(-i2\pi(li/N_x + jm/N_y + kn/N_z)) \tag{10.27}$$

where $i = 0, \ldots N_x - 1$, $j = 0, \ldots N_y - 1$ and $k = 0, \ldots N_z - 1$. The grid points are defined as

$$x_i = i\Delta x + \tfrac{1}{2}\Delta x \tag{10.28}$$

$$y_j = j\Delta y + \tfrac{1}{2}\Delta y \tag{10.29}$$

$$z_k = k\Delta z + \tfrac{1}{2}\Delta z \tag{10.30}$$

The derivative of the wavefunction with respect to x, for example, is now given as

$$\frac{\partial^p \psi}{\partial x^p} = \frac{1}{\sqrt{N}} \sum_{lmn} c_{lmn} \left(\frac{-i2\pi l}{N_x \Delta x}\right)^p \exp(-i2\pi(il/N_x + jm/N_y + kn/N_z)) \tag{10.31}$$

Thus the operator working on the wavefunction can be evaluated by replacing the coefficients c_{lmn} with

$$d_{lmn} = c_{lmn} \left(\frac{-i2\pi l}{N_x \Delta x} \right)^p \tag{10.32}$$

and performing the inverse transform. The Fourier transform algorithm gives the Fourier expansion coefficients [237] such that we should use (for $p = 1$)

$$d_{lmn} = -\frac{i2\pi}{N_x \Delta x} l \text{ for } l < N_x/2 \tag{10.33}$$

$$d_{lmn} = -\frac{i2\pi}{N_x \Delta x} (l - N_x) \text{ for } < l \geq N_x/2 \tag{10.34}$$

with obvious extensions to $p > 1$ and y and z derivatives.

10.3 DIRECT DYNAMICS

So-called ab initio molecular dynamics techniques in which the DFT (usually just in its LDA approximation) is combined with a classical mechanical treatment of the nuclear (ion) motion have been a very popular way of studying condensed matter, i.e., the dynamics of liquids and solids. These techniques may, for instance, be used to study dynamic processes such as binding and atom diffusion [240] at surfaces, and in principle also reactivity at surfaces, without resorting to the usual procedure, namely that of determining the potential energy surface first and then doing the nuclear dynamics afterwards.

The motions of the electrons are adjusted in such a way that the lower or ground state Born-Oppenheimer surface is used to calculate the forces on the nuclei. The method is more time-consuming than standard molecular dynamics methods, which use analytical potential energy surfaces. The advantage is that one does not have to worry about fitting potential energy surfaces, or the global correctness of such a fit, or perform such fitting for each new system studied. At present the methodology has been pushed so far that one is able to treat from about 100 to 1000 atoms, depending on the sophistication of the DFT surface used [241].

The DFT scheme can be put in a molecular dynamics framework by considering the wavefunction as a variable and forming the lagrangian

$$L = \sum_i \mu \langle \dot{\psi}_i \mid \dot{\psi}_i \rangle - E(\{\psi_i\}, \{\mathbf{R}_\alpha\}) + \sum_{ij} \mu_{ij} [\langle \psi_j \mid \psi_i \rangle - \delta_{ij}] \tag{10.35}$$

where μ is a "mass" parameter, $E()$ the energy functional obtained from the Kohn-Sham theory, and the last term with the Lagrange-multipliers μ_{ij} is included to secure that the solutions ψ_i are orthonormal. Application of Lagrange equations of motion

$$\frac{d}{dt}\frac{\partial L}{\partial \dot{\psi}_i^*} = \frac{\partial L}{\partial \psi_i^*} \tag{10.36}$$

give the equations of motion

$$\mu \ddot{\psi}_i = -H_{KS}\psi_i + \sum_j \mu_{ij}\psi_j \tag{10.37}$$

where $H_{KS}\psi_i$ is given by the left-hand side of eq. (10.6). By inserting the expansion of the wavefunction in some suitable basis set, the equations of motion are converted into a set of equations for the expansion coefficients (or in practice any variational parameters of the basis set). The Kohn-Sham hamiltonian then defines the force which drives the wavefunction to a steady-state situation where $\dot{\psi}_i = 0$. The minimum can be found by solving the "equations of motion" using traditional integrators such as the Verlet algorithm [197], but a minimization can also be carried out directly using steepest descent or conjugent gradient methods [198]. In the steepest descent method one searches for the minimum of a function $F(\mathbf{x})$ by following the direction of the negative gradient, i.e.,

$$\mathbf{x}_2 = \mathbf{x}_1 - \frac{\partial F}{\partial \mathbf{x}}|_{\mathbf{x}_1}\, \delta \tag{10.38}$$

where the parameter δ is taken so large that the function still decreases. Sometimes a faster method for searching the minimum is obtained using the conjugate gradient method in which the direction of a given step k in the search for the minimum is obtained as

$$\mathbf{d}^k = \mathbf{g}^k + \frac{\mathbf{g}^k \cdot \mathbf{g}^k}{\mathbf{g}^{k-1} \cdot \mathbf{g}^{k-1}}\mathbf{d}^{k-1} \tag{10.39}$$

where $\mathbf{g}^k$ is the gradient at the k'th step [198].

Once this eigenstate of the electronic hamiltonian is reached, we can use the Hellmann-Feynman theorem for calculating the force on the nuclei (the ions), i.e., if an eigenstate of the electronic hamiltonian is reached, then the force on ion number k is given as

$$\mathbf{f}_k = -\frac{\partial E(\{\mathbf{R}_k\})}{\partial \mathbf{R}_k} \tag{10.40}$$

In order to reach the eigenstate, it is necessary to iterate the above equations in a given time-step, before the nuclei are moved according to Newton's mechanics

$$M_j \ddot{\mathbf{R}}_j = \mathbf{f}_j \tag{10.41}$$

Thus, since the hamiltonian depends parametrically on the nuclear coordinates, we can let the orbital dynamics follow the nuclear motion provided that the nuclear motion is governed by classical mechanics, and the forces on the nuclei can be obtained simultaneously by applying the Hellman-Feynmann theorem. If a global rather than a local minimum (where the forces happen to be zero) is sought, then special techniques known as simulated annealing have to be applied [199].

The scheme in which one solves simultaneously the nuclear and the electronic dynamics is known as the Carr-Parrinello (CP) scheme [239]. The advantage of treating the nuclear and electronic degrees of freedom simultaneously is that ab initio calculations are carried out only for those nuclear configurations needed for obtaining the quantities of interest and that the construction of analytical potential energy surfaces is avoided. The disadvantage of the CP scheme is that the nuclear dynamics are treated using classical mechanics and that the dynamics are followed using just a single electronic adiabatic state. That is, quantum effects such as tunneling, nonadiabatic electronic transitions, wavepacket bifurcation, zero-point vibrational energy, and so on, are either not treated or not treated properly when using classical dynamics. Various attempts have been made to incorporate some of these aspects [268, 242], but at present it is not clear how useful these more general methods will be. Thus a severe testing case for DFT-based theories is how well chemical reactions are treated. The region around the reaction barrier and the barrier itself have to be known with an accuracy of about 0.01 eV or better in order to be of value as predictive tools. The fact that at present we do not have a systematic way of improving the DFT theory may be a severe limitation. From a practical point of view the combination of either transition state theory or the more accurate reaction path theory [156] with DFT calculations would provide a useful first step in the prediction of reaction rates from first principles.

EXERCISE

10.1: Investigate the two limits of eq. (10.16).

11

ELECTRON-HOLE PAIR EXCITATION

In addition to the friction which arises from the excitation of phonons, we must consider the possibility of exciting the electrons in the substrate. Electronic damping of the vibrational motion of adsorbed molecules and atoms has been considered by several groups. Persson and Helsing [249] calculated the vibrational damping rate using an electron-hole pair model based upon a homogeneous electron gas. It was found that the damping rate depended strongly upon the electron density and the distance to the jellium edge. Simple model calculations based on the orthogonality theorem of Anderson [250] were carried out by Gadzuk [251]. Brivio and Grimley considered the nonadiabatic coupling of a chemisorbed atom with the metal using a model similar to the one presented below [252]. Later Metiu et al. considered the role of electron-hole pair excitation in desorption dynamics [256]. Their model included both the phonon and the electron dynamics. Especially the vibrational relaxation of molecules adsorbed at surfaces can be due to electronic damping, the reason being that vibration-to-vibration energy transfer is inefficient due to the large energy mismatch between the vibrational energy of the diatom and that of the phonons. The fact that the damping rate increases with the cluster size also has been used as an indication of electronic damping [263].

Head-Gordon and Tully [264] used the Golden Rule expression

$$\Gamma = \frac{1}{\tau} = \frac{2\pi}{\hbar} \sum_f |H_{if}|^2 \delta(\varepsilon_i - \varepsilon_f) \tag{11.1}$$

to evaluate the lifetime τ of vibrational excited $v = 1$ state of CO molecules at a copper surface. The energies $\varepsilon_{i,f}$ are energies of the initial and final states, i.e., an electronic state of the metal and a vibrational molecular state. The matrix element H_{if} is the transition element between the two states, where the initial electronic state lies below and the final above the Fermi level. From the Hartree-Fock theory for the electrons, the lifetime was estimated to be about 2 ps — in good agreement with the experimentally measured relaxation times [265]. This number was also obtained by Billing [257] using the semiclassical model described below and the Golden Rule expression. Wahnström et al. [258] introduced the frictional force due to electronic excitation in classical dynamical simulations of surface diffusion. The friction constant was estimated for niobium to be about 0.04 10^{14} $\sec^{-1}$. Later Kosloff et al. [260] used the same model for hydrogen transport calculations on a nickel surface. The temperature dependence of the damping rate has been discussed by Persson and Gadzuk [259].

Billing proposed a semiclassical model based on a second quantization approach for the electronic degrees of freedom [262]. The various models proposed have given conflicting results concerning the role of electron-hole pair excitation from large effects [253] over small effects for atoms as He [254] to negligible effects for He on a Cu-surface [255]. The relative role of friction due to phonon excitation as compared to electron-hole pair excitation also has been debated [256]. Thus at present it appears that the situation regarding electron-hole pair excitation is not as clear as could be desired.

In order to evaluate the influence of phonon versus electron friction it is necessary to have models which include both aspects. Although many models for electron-hole pair excitation are available, most of these do not also consider the phonon excitation, and they are therefore not able to discriminate between the two processes. The model which is proposed by Metiu and coworkers [256] as well as Billing [262] assumes that the effective charge on the incoming molecule or adsorbate through Coulomb interaction with the metal-electrons excites these from levels below to levels above the Fermi level. Below we describe the semiclassical model proposed by Billing. This model has a collision-oriented aspect and treats the excitation processes to infinite order.

One obvious question is whether the nuclear and electronic motion can be separated in the fashion which is done in most models for molecule surface scattering and also in the above-mentioned treatment of electron-hole pair excitation. The traditional approach is to invoke a Born-Oppenheimer approximation, i.e., one defines adiabatic potential energy surfaces on which the nuclear dynamics is solved — either quantally or classically. In the Born-Oppenheimer picture the electrons have had enough time to readjust to the nuclear positions. Thus the nuclei are assumed to move infinitely slowly. For finite speed, nonadiabatic corrections therefore have to be introduced. Thus, before comparison with experimental data is carried out we have to consider whether nonadiabatic processes are important. Two types of nonadiabatic processes are possible — one is nonadiabatic transitions in the gas phase from the lower adiabatic to the upper surface (as discussed in Chapter 4). The other is the nonadiabatic excitation of electrons in the metal through electron-hole pair excitation.

The total hamiltonian for the crystal or crystal + adatom can be written as

$$H = T_e + T_I + V_{eI}(\mathbf{r}, \mathbf{R}) \tag{11.2}$$

where T_e and T_I are kinetic energies of the electrons and ions respectively and V_{eI} the interaction energy, i.e., for instance, the Coulomb interaction or pseudopotentials between valence electrons and ionic cores, core-core, and electron-electron interactions. $\mathbf{r}$ denotes a set of electronic and $\mathbf{R}$ a set of ion coordinates. Adiabatic electronic states are obtained by solving the electronic problem

$$(T_e + V_{eI}(\mathbf{r}, \mathbf{R}))\phi_n(\mathbf{r}; \mathbf{R}) = E_n(\mathbf{R})\phi_n(\mathbf{r}; \mathbf{R}) \tag{11.3}$$

and then an expansion is made in these adiabatic states

$$\Psi(\mathbf{r}, \mathbf{R}) = \sum_n \Psi_n(\mathbf{R})\phi_n(\mathbf{r}; \mathbf{R}) \tag{11.4}$$

and insertion in the Schrödinger equation gives a set of coupled equations for the nuclear motion

$$[T_I + E_n(\mathbf{R}) - E]\psi_n(\mathbf{R}) - \sum_m A_{nm}(\mathbf{R})\psi_m(\mathbf{R}) = 0 \tag{11.5}$$

where A_{nm} are the so-called nonadiabatic coupling elements

$$A_{nm} = \frac{\hbar^2}{M}\int d\mathbf{r}\phi_n^*(\mathbf{r}; \mathbf{R})\nabla_{\mathbf{R}}\phi_m(\mathbf{r}; \mathbf{R})\nabla_{\mathbf{R}} + \frac{\hbar^2}{2M}\int d\mathbf{r}\phi_n^*(\mathbf{r}; \mathbf{R})\nabla_{\mathbf{R}}^2\phi_m(\mathbf{r}, \mathbf{R}) \tag{11.6}$$

where for simplicity we have assumed that all atoms have the mass M. Obviously there will be a continuum of electronic adiabatic states, and the solution scheme must be simplified in one way or another, as discussed below, for instance, by introducing one-electron models for electrons moving in effective core potentials and an approximate treatment of the problem expressed, for example, in the second quantization language. In order to simplify the description further we shall furthermore switch below to a time-dependent theory in which the nuclear motion of the atom is assumed to follow a classical trajectory.

11.1 ONE-ELECTRON MODELS

The periodicity of the metal makes the potential periodic in the lattice surface coordinates, such that

$$V(\mathbf{R}, z) = V(\mathbf{R} + \mathbf{R}_g, z) \tag{11.7}$$

where $\mathbf{R}_g$ is a vector in the (X, Y) plane reflecting the periodicity of the crystal. This periodicity can be utilized when searching for simple models for describing the electronic excitation.

In order to facilitate the discussion one often introduces a one-electron model for the motion of the electrons in the metal, i.e.,

$$\left(-\frac{\hbar^2}{2m}\nabla_r^2 + U(r)\right)\phi_n(r) = E_n\phi_n(r) \tag{11.8}$$

where m is the mass of the electron and $U(r)$ is some effective potential. The simplest one-electron model assumes a potential of the type

$$U(x, y, z) = -U_0 \text{ constant for } 0 < x < L$$
$$U(x, y, z) = 0 \text{ otherwise}$$

Other models include the corrugation of the surface. Thus one can introduce a Bloch-potential as

$$U(x) = 2U_0\cos(2\pi x/a) \tag{11.9}$$

where a is the distance between the ions. The Bloch potential model is readily extended to two or three dimensions and the Bloch wavefunctions can be obtained by linear combination of atomic orbitals (LCAO) as [266]

$$\psi_k(\mathbf{r}) = \sum_{n_1 n_2 n_3} \exp(i\mathbf{k}\cdot\mathbf{An})\phi(\mathbf{r}-\mathbf{An}) \tag{11.10}$$

where $\mathbf{n} = (n_1, n_2, n_3)$, $\phi(\mathbf{r})$ is an atomic orbital, $\mathbf{A}$ the matrix containing the lattice vectors (see Chapter 2), and the summation runs over lattice points. The basis functions can easily be shown to be periodic, i.e., $\psi_\mathbf{k}(\mathbf{r}+\mathbf{An}) = \exp(i\mathbf{k}\cdot\mathbf{An})\psi_\mathbf{k}(\mathbf{r})$, and can be used for more elaborate electronic band structure calculations. This is, however, not the subject of the present monograph and we shall therefore return to the simple one-dimensional (1D) picture.

With the potential (11.9), the solution to the Schrödinger equation (SE) (11.8) can be carried out. We find that

$$\phi_k = \sum_G \exp(i(k-G)x)C(k-G) \tag{11.11}$$

Inserting this expansion in the time-independent SE we get an infinite matrix equation of the form

$$\begin{pmatrix} \lambda_{k-2G}-\varepsilon_k & U_0 & 0 & 0 \\ U_0 & \lambda_{k-G}-\varepsilon_k & U_0 & 0 \\ 0 & U_0 & \lambda_k-\varepsilon_k & U_0 \\ 0 & 0 & U_0 & \lambda_{k+G}-\varepsilon_k \end{pmatrix}\begin{pmatrix} C(k-2G) \\ C(k-G) \\ C(k) \\ C(k+G) \end{pmatrix} = \begin{pmatrix} 0 \\ 0 \\ 0 \\ 0 \end{pmatrix} \tag{11.12}$$

where $\lambda_k = \hbar^2 k^2/2\,m$. The simplest solution of this infinite set of equations is obtained using a two-band model, where

$$\begin{pmatrix} \lambda_{k-G} - \varepsilon_k & U_0 \\ U_0 & \lambda_k - \varepsilon_k \end{pmatrix} \begin{pmatrix} C(k-G) \\ C(k) \end{pmatrix} = \begin{pmatrix} 0 \\ 0 \end{pmatrix} \tag{11.13}$$

from which we can determine the eigenvalues

$$\varepsilon_k^{\pm} = \frac{1}{2}\left(\lambda_k + \lambda_{k-G} \pm (\lambda_k - \lambda_{k-G})\sqrt{1 + \frac{4U_0^2}{(\lambda_k - \lambda_{k-G})^2}} \right) \tag{11.14}$$

and eigenvectors

$$\begin{pmatrix} C(k-G) \\ C(k) \end{pmatrix} = \begin{pmatrix} \pm\sqrt{1 - C(k)^2} \\ \pm U_0 / \sqrt{U_0^2 + (\lambda_k - \varepsilon_k^{\pm})^2} \end{pmatrix} \tag{11.15}$$

where $G = 2\pi/a$. The energy as a function of the wavevector k for this two-band model is shown in Figure 11.1. Notice the energy band gap. Excitation of electrons in solids which have only the lowest band occupied requires excitation over this gap. Such solids are known as insulators. However, for metals, the Fermi energy is such that the second band is occupied and electron-hole (e-h) pair excitation comes about as a result of excitation of electrons with energies near the Fermi level to above this level.

More general expressions use a two-dimensional (2D) wavefunction and more than one band, i.e., for a $(2n_b + 1)$ band model with Bloch wavefunctions in

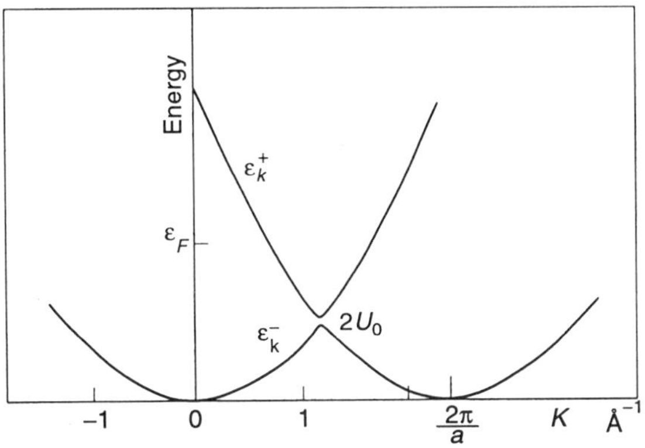

FIGURE 11.1 Band structure obtained with a simple 2-band model. G_1 is the reciprocal lattice vector $2\pi/a$, where a is the lattice constant, ε_F is the Fermi energy, and $2U_0$ the band gap.

both the x and y direction we have

$$\phi_{\mathbf{k},n} = \sum_{j=1,2} \sum_{l=-n_b}^{n_b} C_{\mathbf{k}+l\mathbf{b}_j,n} \exp(i(\mathbf{k} + l\mathbf{b}_j) \cdot \mathbf{r})) \tag{11.16}$$

where $\mathbf{b}_j$ are the reciprocal lattice vectors and $\mathbf{r}$ the position coordinate of the electron [267]

The electron excitation process is caused by the interaction with the incoming atom or molecule. Since we have seen that some charge transfer takes place from the metal to the incoming atom/molecule the simplest model assumes that this charged particle can induce the e-h excitation through its interaction with the electrons. In this approach the e-h processes are viewed as additional inelastic (nonadiabatic) processes taking place "on top" of the adiabatic electronic adjustment connected with the charge transfer and the Born-Oppenheimer construction of the adiabatic surfaces used for the nuclear motion. In order to facilitate the theoretical treatment of these inelastic processes, it is convenient to introduce the concept of second quantization.

11.2 SECOND QUANTIZATION APPROACH

We introduce the creation/annihilation operators for electrons such that c_i^+ creates an electron in state i, i.e.,

$$c_i^+|n_1, n_2, \ldots n_i, \ldots >= \sqrt{1 - n_i}(-1)^{\sum_{j<i} n_j}|n_1, n_2, \ldots n_i + 1, \ldots > \tag{11.17}$$

Thus we see that the result is zero if the state is occupied, i.e., if $n_i = 1$. The phase factor has to do with the requirement of antisymmetry of the wavefunction for Fermi-ions, i.e., the ordering of the states plays an important role. Likewise we have

$$c_i|n_1, n_2, \ldots n_i, \ldots >= \sqrt{n_i}(-1)^{\sum_{j<i} n_j}|n_1, n_2, \ldots n_i - 1, \ldots > \tag{11.18}$$

The Hamiltonian for the molecule-surface electron system is expressed as

$$H_e = H_0 + \sum_{i=1}^{N} H_i \tag{11.19}$$

where H_0 is the hamiltonian for the isolated electronic system consisting of N electrons described by the one-electron wavefunctions $\phi_i(r)$. H_i is the hamiltonian for the interaction of the molecule and the ith electron in the solid. The interaction between a particle A in the gas phase and an electron in the solid can for instance be modeled using the Fermi-Thomas model, i.e., by a screened

Coulomb interaction

$$H_i = \frac{\delta e^2 \exp(-k_c \overline{X})}{\overline{X}} \tag{11.20}$$

where δe is the excess charge on the atom, k_c a screening coefficient (see eq. (9.40)), and

$$\overline{X} = |\mathbf{R} - \mathbf{r}_i| \tag{11.21}$$

where $\mathbf{R}$ is the coordinate-vector of the atom and $\mathbf{r}_i$ that of the electron i in the solid. This model appears to be feasible for atom/molecule-metal scattering, since electronic charge is being transferred from the metal to the atom/molecule during the approach toward the surface. The matrix element for the transition from state i to state j is

$$\int d\mathbf{r}\phi_i(\mathbf{r})H(\mathbf{r}, \mathbf{R}(t))\phi_j(\mathbf{r}) = H_{ij}(\mathbf{R}(t)) \tag{11.22}$$

where the atom/molecule trajectory $R(t)$ and the one-electron wavefunctions $\phi_i(\mathbf{r})$ have been introduced. The one-electron wavefunctions are obtained either through a simple model such as the Bloch model, or more rigorously by solving the Kohn-Sham equations.

In second quantization language we have the following hamiltonian [247, 156, 248] for the problem

$$H_e = \sum_i \varepsilon_i c_i^+ c_i + \sum_{ij} H_{ij}(t) c_i^+ c_j \tag{11.23}$$

where ε_i is the energy of electronic state i. The evolution of the electronic system under the external time-dependent force is governed by the evolution operator U_e where

$$i\hbar \frac{d}{dt} U_e(t, t_0) = H_e(t) U_e(t, t_0) \tag{11.24}$$

This equation is solved using

$$U_e = U_0 U_I \tag{11.25}$$

where U_0 is the solution obtained with just the first term in eq. (11.23) and hence

$$i\hbar \frac{d}{dt} U_I = U_0^+ \sum_{ij} H_{ij}(t) c_i^+ c_j U_0 U_I = \sum_{ij} B_{ij} c_i^+ c_j U_I \tag{11.26}$$

where B_{ij} is given below; see (eq. (11.44)). Below we also introduce $\mathbf{R}$ as the matrix representation of the operator U_I.

In order to evaluate the matrix elements H_{ij}, it is convenient to expand the interaction potential as

$$\frac{\delta(z)e^2}{X}\exp(-k_c\overline{X}) = \frac{2e^2\delta(z)}{a}\sum_m \exp(imx/a)K_0\left(z\sqrt{(m/a)^2+k_c^2}\right) \tag{11.27}$$

where K_0 is a modified Bessel function and where we have used the fact that $\overline{X}^2 = z^2 + x^2$ (see Fig. 11.2), z is the distance from the atom/molecule to the jellium edge, and a is the lattice constant. The matrix elements between the one-electron wavefunctions and the interaction potential are now obtained as

$$\langle\phi_k|\exp(imx/a)|\phi_{k'}\rangle = \begin{cases} C(k)C(k') + C(k-G)C(k'-G) & k' = k - m/a \\ C(k-G)C(k') & k' = k - G - m/a \\ C(k'-G)C(k) & k' = k + G - m/a \end{cases} \tag{11.28}$$

that is, we have a set of selection rules for the electronic transitions from states below to states above the Fermi level. The values of $C(k)$ are obtained from eqs. (11.15)

11.2.1 Two-Dimensional Bloch Model

If a 2D Bloch model is assumed, we can proceed in a manner similar to that shown above and write [179]

$$\frac{e^2\delta}{\overline{X}}\exp(-k_c\overline{X}) = \sum_{nm} A_{nm}(z)\exp(i(mx/a + ny/a)) \tag{11.29}$$

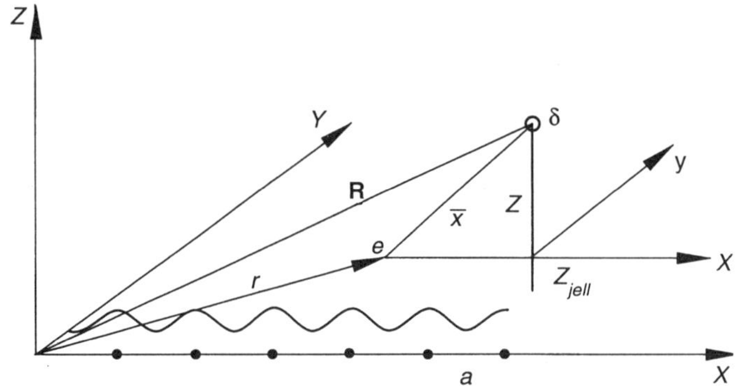

FIGURE 11.2 The excess charge δ on an atom or molecule interacts with an electron e through a screened Coulomb potential. The position vector of the atom is $\mathbf{R} = (X, Y, Z)$. The distance $\overline{X} = \sqrt{x^2 + y^2 + z^2}$ where (x, y) is the position of the electron moving in a 2D Bloch potential and z the distance to the "jellium" edge $z = Z - z_{jell}$.

where $\overline{X}^2 = x^2 + y^2 + z^2$ and where for the lattice we have assumed $a = b$. The Fourier coefficients are

$$A_{nm}(z) = \frac{e^2\delta}{2\pi a}\frac{1}{\sqrt{n^2 + m^2 + (ak_c)^2}}\exp\left(-z\sqrt{k_c^2 + (n/a)^2 + (m/a)^2}\right) \tag{11.30}$$

where a is the lattice constant and z the distance from the excess charge to the jellium edge (Fig. 11.2). The transition matrix elements

$$\langle\phi_{\mathbf{k},n}|\exp(imx/a + iny/a)|\phi_{\mathbf{k}',n'}\rangle \tag{11.31}$$

can also be evaluated analytically [267] (see Table 11.1). Thus with 2D Bloch functions from eq. (11.16) we get 13 selection rules instead of the three obtained with the 1D Bloch functions. The value of $\mathbf{k}'$ is furthermore restricted to lie in the first Brillouin zone $(-\pi/a; +\pi/a)$. The reason for this is that in the Bloch wavefunction we may replace k with $k + (2\pi n/a)$ without changing the form of the function. It is therefore possible to restrict the values of k to an interval of length $2\pi/a$. This interval is conveniently taken symmetrically around 0, i.e., from $-\pi/a$ to $+\pi/a$.

11.2.2 Density of States

Transitions from the initial state k to the final k' are induced by the matrix elements (11.31). The wavevector k of the initial state can take values up to the

TABLE 11.1 Matrix Elements for Electronic Transitions Obtained Using the 2D Bloch Model

Matrix Element	Selection Rule
$C(\mathbf{k}-\mathbf{b}_1)C(\mathbf{k}'-\mathbf{b}_1) + C(\mathbf{k}-\mathbf{b}_2)$ $C(\mathbf{k}'-\mathbf{b}_2) + C(\mathbf{k})C(\mathbf{k}')+$ $C(\mathbf{k}+\mathbf{b}_2)C(\mathbf{k}'+\mathbf{b}_2) + C(\mathbf{k}+\mathbf{b}_1)C(\mathbf{k}'+\mathbf{b}_1)$	$\mathbf{k}' = \mathbf{k} - (m/a, n/a)$
$C(\mathbf{k}-\mathbf{b}_2)C(\mathbf{k}') + C(\mathbf{k})C(\mathbf{k}'+\mathbf{b}_2)$	$\mathbf{k}' = \mathbf{k} - \mathbf{b}_2 - (m/a, n/a)$
$C(\mathbf{k}+\mathbf{b}_2)C(\mathbf{k}') + C(\mathbf{k})C(\mathbf{k}'-\mathbf{b}_2)$	$\mathbf{k}' = \mathbf{k} + \mathbf{b}_2 - (m/a, n/a)$
$C(\mathbf{k}-\mathbf{b}_1)C(\mathbf{k}') + C(\mathbf{k})C(\mathbf{k}'+\mathbf{b}_1)$	$\mathbf{k}' = \mathbf{k} - \mathbf{b}_1 - (m/a, n/a)$
$C(\mathbf{k}+\mathbf{b}_1)C(\mathbf{k}') + C(\mathbf{k})C(\mathbf{k}'-\mathbf{b}_1)$	$\mathbf{k}' = \mathbf{k} + \mathbf{b}_1 - (m/a, n/a)$
$C(\mathbf{k}-\mathbf{b}_2)C(\mathbf{k}'+\mathbf{b}_2)$	$\mathbf{k}' = \mathbf{k} - 2\mathbf{b}_2 - (m/a, n/a)$
$C(\mathbf{k}+\mathbf{b}_2)C(\mathbf{k}'-\mathbf{b}_2)$	$\mathbf{k}' = \mathbf{k} + 2\mathbf{b}_2 - (m/a, n/a)$
$C(\mathbf{k}-\mathbf{b}_2)C(\mathbf{k}'-\mathbf{b}_1) + C(\mathbf{k}+\mathbf{b}_1)C(\mathbf{k}'+\mathbf{b}_2)$	$\mathbf{k}' = \mathbf{k} + \mathbf{b}_1 - \mathbf{b}_2 - (m/a, n/a)$
$C(\mathbf{k}-\mathbf{b}_2)C(\mathbf{k}'-\mathbf{b}_1) + C(\mathbf{k}-\mathbf{b}_1)C(\mathbf{k}'+\mathbf{b}_2)$	$\mathbf{k}' = \mathbf{k} - \mathbf{b}_1 - \mathbf{b}_2 - (m/a, n/a)$
$C(\mathbf{k}+\mathbf{b}_2)C(\mathbf{k}'-\mathbf{b}_1) + C(\mathbf{k}+\mathbf{b}_1)C(\mathbf{k}'-\mathbf{b}_2)$	$\mathbf{k}' = \mathbf{k} + \mathbf{b}_1 + \mathbf{b}_2 - (m/a, n/a)$
$C(\mathbf{k}+\mathbf{b}_2)C(\mathbf{k}'+\mathbf{b}_1) + C(\mathbf{k}-\mathbf{b}_1)C(\mathbf{k}'-\mathbf{b}_2)$	$\mathbf{k}' = \mathbf{k} + \mathbf{b}_2 - \mathbf{b}_1 - (m/a, n/a)$
$C(\mathbf{k}-\mathbf{b}_1)C(\mathbf{k}'-\mathbf{b}_1)$	$\mathbf{k}' = \mathbf{k} - 2\mathbf{b}_1 - (m/a, n/a)$
$C(\mathbf{k}+\mathbf{b}_1)C(\mathbf{k}'-\mathbf{b}_1)$	$\mathbf{k}' = \mathbf{k} + 2\mathbf{b}_1 - (m/a, n/a)$

The band index has been omitted for simplicity.

energy defined by the Fermi-energy ε_F using eq. (11.14). Instead of summing over initial states k we can replace the sum with an integral

$$\sum_{k=1}^{F} \rightarrow \int_0^{\varepsilon_F} d\varepsilon \rho(\varepsilon) h(\varepsilon) \tag{11.32}$$

where $\rho(\varepsilon)$ is the density of states and $h(\varepsilon)$ is zero for energies in the band gap between E_{min} and E_{max}. For a free-electron gas the density of states $\rho(\varepsilon)$ is given by eq. (11.6) with $\rho(\varepsilon) = 2\,\mathcal{N}(\varepsilon)$ and $\varepsilon = \hbar^2 k^2/2\,m$, i.e.,

$$\rho(\varepsilon) = \frac{V}{\pi^2}\frac{mk}{\hbar^2} \tag{11.33}$$

where $V = (3\pi^2 N)/k_F^3$, and N is the number of electrons in the volume V. But the density of states could be taken either from experimental data or from more extensive band structure calculations. In this manner it is possible to incorporate the effect that the density of states changes near the surface as compared with the bulk. This so-called band narrowing at the surface can be incorporated using simple models such as the tight binding model (equivalent to the Hückel model in chemistry). In this model the wavefunction is expanded in localized atomic orbitals and the Schrödinger equation for the problem is

$$\mathbf{Ha} = E\mathbf{a} \tag{11.34}$$

Here the vector $\mathbf{a}$ contains the expansion coefficients and the matrix elements of H are over the atomic orbitals $H_{ij} = \langle i|H|j\rangle$. The local charge density $\rho(\mathbf{r})$ can be obtained as

$$\rho(\mathbf{r}) = \sum_i |\psi_i(\mathbf{r})|^2 \tag{11.35}$$

where the sum is over occupied orbitals and the orbitals found by solving the above eigenvalue problem.

The density of states around $\mathbf{r}$ at a fixed energy E is defined through the equation

$$\rho(\mathbf{r}, E) = \sum_i |\psi_i(\mathbf{r})|^2 \delta(E - E_i) \tag{11.36}$$

If we assume that the electronic wavefunction is periodic in the (x, y) direction of the lattice and has an infinite barrier at $z = 0$ and at L_z in a metal slab with thickness L_z, the wave function is simply

$$\psi_{k_x k_y k_z} = \sqrt{2/AL_z}\exp(i\mathbf{K}\cdot\mathbf{R})\sin(k_z z) \tag{11.37}$$

where $\mathbf{R} = (x, y)$, $\mathbf{K} = (k_x \cdot k_y)$ and A the surface area. With this wavefunction we get the density of states

$$\rho(z, E) = \rho(E)(1 - \sin(az)/(az)) \tag{11.38}$$

where $\rho(E) = (1/2\pi^2)\ (2m/\hbar^2)^{3/2}\sqrt{E}$ and $a = 2\sqrt{2mE}/\hbar$. The total density of states can be obtained by integrating over energies from zero to the Fermi energy ε_F [261]

$$\rho(z) = \int_0^{\varepsilon_F} \rho(E, z) dE = \frac{k_F^3}{3\pi^2}\left(1 + \frac{3\cos\bar{z}}{\bar{z}^2} - \frac{3\sin(\bar{z})}{\bar{z}^3}\right) \tag{11.39}$$

where $\bar{z} = 2k_F z$, k_F the Fermi wavevector, and the density of a Fermi gas given as $\rho_0 = k_F^3/3\pi^2$ (see Chapter 9). This model shows oscillations in the density near the surface edge. The position of the potential edge z_d obtained with this model is given by requiring that the positive charge form a uniform background

$$\rho^+(z) = \rho^\theta(z - z_d) \tag{11.40}$$

where $\theta(x)$ is Heaviside function (unity for negative and zero for positive arguments) balancing the negative charge. Thus by requiring that

$$\int (\rho^-(z) - \rho^+(z))\, dz = 0 \tag{11.41}$$

we get using (11.39) for $\rho^-(z)$ that $z_d = 3\pi/8k_F$.

11.2.3 The R-Matrix Problem

Equations (11.24) can be solved using the second quantization expression for H_e given by (11.23). They are identical to the eqs. (8.76) for the phonon VV problem, but with fermion operators $c_k^+ c_l$ instead of boson operators $a_k^+ a_l$. However, the operators $c_k^+ c_l$ also form a closed set with respect to commutations and the algebraic problem is therefore identical to the one for bosons, i.e., we have

$$U_I = \exp\left(\sum_{ij} \alpha_{ij} c_i^+ c_j\right) \tag{11.42}$$

where the functions α_{ij} can be related to the functions Q_{kl}. These functions are obtained from the solution to the matrix equation

$$i\hbar\frac{d}{dt}\mathbf{R} = \mathbf{BR} \tag{11.43}$$

where $\mathbf{Q} = \mathbf{R} - \mathbf{I}$ (the unit matrix) and

$$B_{ij} = H_{ij} \exp\left(\frac{i}{\hbar}(\varepsilon_i - \varepsilon_j)t\right) \tag{11.44}$$

where i and j label the one-electron states. An approximate but unitary solution to the above equations is

$$\mathbf{R}(t, t_0) = \exp\left(\frac{1}{i\hbar}\int_{t_0}^{t} dt'\mathbf{B}(t')\right)\mathbf{R}(t_0, t_0) \tag{11.45}$$

with $\mathbf{R}(t_0, t_0) = \mathbf{I}$ and where $B_{ij} = B^*_{ji}$. The exponential function can be evaluated if we make the assumption that a given state i (below the Fermi level) couples to a set of states $j(i)$ above the Fermi level F through the selection rules posed by the above matrix elements (Fig. 11.3). The parentheses (i) indicates that these levels are the ones to which level i is coupled. Thus the B-matrix is assumed to take the following form for level i

$$\begin{matrix}\text{Below} \\ \\ \\ \text{Above} \\ \\ \\ \end{matrix}\begin{bmatrix} 0 & x & x & x & x & x \\ x & 0 & 0 & 0 & 0 & 0 \\ x & 0 & 0 & 0 & 0 & 0 \\ x & 0 & 0 & 0 & 0 & 0 \\ x & 0 & 0 & 0 & 0 & 0 \\ x & 0 & 0 & 0 & 0 & 0 \end{bmatrix} \tag{11.46}$$

where x denotes the nonzero matrix elements (here just 5). For another state, say "$i + 1$," we get a similar coupling matrix. Since the number of m-terms we include in the Fourier expansion (11.27) is much smaller than the number of ϕ_j-states, we can also assume that the set of states $j(i)$ is different from the set $j(i + 1)$, and so on. Thus the individual coupling matrices can be taken to be independent.

This also implies that states below or above the Fermi level are coupled only through the higher-order terms obtained when exponentiating the matrix. For

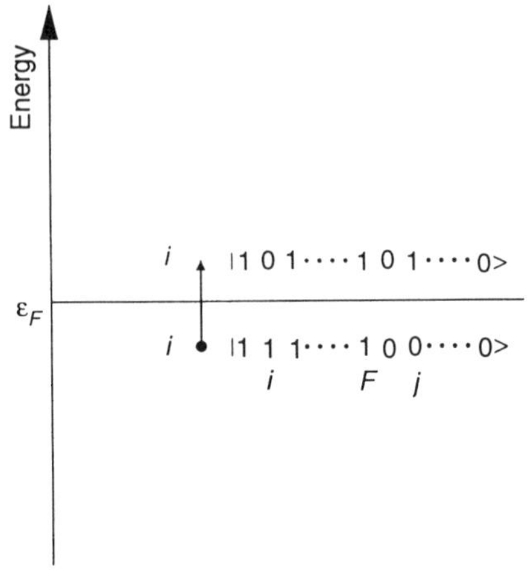

FIGURE 11.3 A state below the Fermi level is excited to a state above. The matrix element connected to this transition is H_{ij} and it is induced by the operator $c_i c_j^+$.

each element i below the Fermi level we have a similar coupling matrix. If we square such a matrix of the type in (11.46) the structure becomes

$$\begin{bmatrix} & 0 & 0 & 0 & 0 & 0 \\ 0 & x & x & x & x & x \\ 0 & x & x & x & x & x \\ 0 & x & x & x & x & x \\ 0 & x & x & x & x & x \\ 0 & x & x & x & x & x \end{bmatrix} \tag{11.47}$$

where again x denotes an element different from zero.

Considering now the exponentiation of such a matrix, we have

$$\exp(-i\mathbf{A}^{(1)}) = \sum_{n=0}^{\infty} \frac{1}{n!}(-i\mathbf{A}^{(1)})^n \tag{11.48}$$

Thus we have for $\mathbf{A}^{(1)}\mathbf{A}^{(1)}$ the element

$$A_{kk}^{(2)} = \sum_{l} |A_{kl}^{(1)}|^2 \tag{11.49}$$

$$A_{ll'}^{(2)} = \sum_{k} A_{lk}^{(1)} A_{kl'}^{(1)} \tag{11.50}$$

where k is a state below the Fermi level $k < F$, whereas $l, l' > F$, and

$$A_{kl}^{(1)} = \frac{1}{\hbar}\int_{t_0}^{t} dt' H_{kl} \exp\left(\frac{i}{\hbar}(\varepsilon_k - \varepsilon_l)t'\right) \tag{11.51}$$

In a similar manner we have $(\mathbf{A}^{(1)})^3 = \mathbf{A}^{(3)}$ with

$$A_{kl}^{(3)} = A_{kk}^{(2)} A_{kl}^{(1)} \tag{11.52}$$

$$A_{lk}^{(3)} = \sum_{l'} A_{l'k}^{(1)} A_{kl'}^{(1)} A_{lk}^{(1)} = A_{kk}^{(2)} A_{lk}^{(1)} \tag{11.53}$$

and so on. Here we have also used the fact that $A_{kl}^{(1)} = A_{lk}^{(1)*}$. Thus the odd terms sum to

$$R_{kl} = -iA_{kl}^{(1)} + \frac{(-i)^3}{3!} A_{kk}^{(2)} A_{kl}^{(1)} + \frac{(-i)^5}{5!} (A_{kk}^{(2)})^2 A_{kl}^{(1)} + \cdots = -i \frac{A_{kl}^{(1)}}{\sqrt{A_{kk}^{(2)}}} \sin\left(\sqrt{A_{kk}^{(2)}}\right) \tag{11.54}$$

where $k \neq l$ and the even terms sum to

$$R_{kk} = 1 + \frac{i^2}{2!} A_{kk}^{(2)} + \frac{i^4}{4!}\left(A_{kk}^{(2)}\right)^2 + \cdots = \cos\left(\sqrt{A_{kk}^{(2)}}\right) \quad \text{and} \tag{11.55}$$

$$R_{ll'} = \frac{A_{ll'}^{(2)}}{A_{kk}^{(2)}} \cos\left(\sqrt{A_{kk}^{(2)}}\right) \tag{11.56}$$

where $l \neq l'$, and, as mentioned above, we also have $k < F$ and $l, l' > F$. For the state k below the Fermi level we have

$$|R_{kk}|^2 + \sum_{l \neq k} |R_{kl}|^2 = \cos^2\left(\sqrt{A_{kk}^{(2)}}\right) + \frac{\sum_l |A_{kl}^{(1)}|^2}{A_{kk}^{(2)}} \sin^2\left(\sqrt{A_{kk}^{(2)}}\right) = 1 \quad (11.57)$$

which confirms the expected unitarity. Thus for each state $k < F$ we have an expression similar to the one given above.

11.3 THE EFFECTIVE POTENTIAL

We can now evaluate the effective mean field potential used to couple the electronic excitation to the motion of the incoming molecule. The effective hamiltonian is defined as the expectation value, i.e.,

$$H_{eff}^{elho} = \langle \Psi | H_e | \Psi \rangle \quad (11.58)$$

where

$$\Psi(t) = U_{el}(t, t_0)\Psi(t_0) \quad (11.59)$$

and where $\Psi(t_0)$ is the initial wavefunction. The evolution operator is expressed as $U_{el} = U_0 U_I$, where U_0 is the evolution operator under the hamiltonian $H_0 = \sum_k \varepsilon_k c_k^+ c_k$. Using (11.23) we get

$$H_{eff}^{elho} = \sum_k \varepsilon_k n_k + V_{eff}^{elho} \quad (11.60)$$

where V_{eff}^{elho} is an effective potential

$$V_{eff}^{elho} = \sum_{kl} (Q_{kl}^* n_l + Q_{lk} n_k) B_{kl} + \sum_{kli} Q_{ki}^* Q_{li} B_{kl} n_i \quad (11.61)$$

where we have used [262, 156]

$$U_I^+ c_k^+ U_I = c_k^+ + \sum_j Q_{kj}^* c_j^+ \quad (11.62)$$

$$U_I^+ c_l U_I = c_l + \sum_j Q_{lj} c_j \quad (11.63)$$

$$\langle \Psi(t_0) | c_k^+ c_l | \Psi(t_0) \rangle = \delta_{kl} n_k \quad (11.64)$$

and the fact that $U_0^+ H_{ij} c_i^+ c_j U_0 = B_{ij} c_i^+ c_j$.

The interaction potential is, as mentioned, given by eq. (11.61). In order to simplify this expression we notice that n_l denotes the occupation number of state

l, i.e., at temperatures well below the Fermi temperature we have that $n_l = 1$ for $l < F$ and $n_l = 0$ for $l > F$. Since the number of electrons being excited during the collision process is modest compared with the total number of electrons, we will assume that we can use these values during the collision process. Thus we get

$$V_{eff}^{elho} = \sum_{l=1}^{F}\sum_{k=1}^{N} B_{kl}Q_{kl}^{*} + \sum_{k=1}^{F}\sum_{l=1}^{N} Q_{lk}B_{kl} + \sum_{kl}\sum_{i=1}^{F} Q_{ki}^{*}Q_{li}B_{kl} \tag{11.65}$$

We now introduce the approximation that the potential V_{kl} couples states below F and states above F, i.e., if k denotes a state below the Fermi-level then l denotes one above and vice-versa. Due to eq. (11.44) the same holds for B_{kl}. Thus we get

$$\begin{aligned} V_{eff}^{elho} &= \sum_{l=1}^{F}\sum_{k=F+1}^{N} B_{kl}Q_{kl}^{*} + \sum_{k=1}^{F}\sum_{l=F+1}^{N} Q_{lk}B_{kl} \\ &\quad + \sum_{i=1}^{F}\sum_{k=1}^{F}\sum_{l=F+1}^{N} Q_{ki}^{*}Q_{li}B_{kl} + \sum_{i=1}^{F}\sum_{l=1}^{F}\sum_{k=F+1}^{N} Q_{ki}^{*}Q_{li}B_{kl} \\ &= \sum_{k=1}^{F}\sum_{l=F+1}^{N} B_{kl}\left(Q_{lk} + \sum_{i=1}^{F} Q_{ki}^{*}Q_{li}\right) \\ &\quad + \sum_{l=1}^{F}\sum_{k=F+1}^{N} B_{kl}\left(Q_{kl}^{*} + \sum_{i=1}^{F} Q_{ki}^{*}Q_{li}\right) \end{aligned} \tag{11.66}$$

where N is the highest state allowed energetically. Using now the fact that

$$\sum_{i=1}^{F} Q_{ki}^{*}Q_{li} = Q_{kk}^{*}Q_{lk} + \sum_{i\neq k} Q_{ki}^{*}Q_{li} \tag{11.67}$$

and that $Q_{ki} = R_{ki} - \delta_{ki}$, we can obtain

$$\begin{aligned} V_{eff}^{elho} &= \sum_{k=1}^{F}\sum_{l=F+1}^{N} (B_{kl}R_{kk}^{*}Q_{lk} + B_{lk}R_{kk}Q_{lk}^{*}) \\ &\quad + \sum_{k=1}^{F}\sum_{l=F+1}^{N}\sum_{i\neq k}^{F} (B_{kl}Q_{ki}^{*}Q_{li} + B_{lk}Q_{li}^{*}Q_{ki}) \end{aligned} \tag{11.68}$$

We notice that the quantity Q_{ki} in the last sum couples two states below the Fermi level, i.e., the coupling we previously neglected. Thus it is consistent also to drop this term here. In addition, we also see that the last term is of second order in Q_{kl} — a small quantity which to first order is given by

$$Q_{ij} \sim \frac{1}{i\hbar} \int_{t_0}^{t} B_{ij}(t')\,dt' \tag{11.69}$$

Thus for several reasons we can expect that the second term in eq. (11.68) is small. If we neglect it we are left with just the expression

$$V_{eff} = 2\mathrm{Re}\left(\sum_{k=1}^{F} \sum_{l=F+1}^{N} B_{kl} R_{kk}^{*} Q_{lk} \right) \tag{11.70}$$

where we have used the fact that $B_{lk} = B_{kl}^{*}$. If we solve the equations for R_{kl} in the so-called time-dependent sudden limit, i.e., we use the fact that (see section 11.2)

$$R_{kk} = \cos\left(\sqrt{A_{kk}^{(2)}}\right) \tag{11.71}$$

$$Q_{lk} = -iA_{kl}^{(1)*} (A_{kk}^{(2)})^{-1/2} \sin\left(\sqrt{A_{kk}^{(2)}}\right) \quad k < F \text{ and } l > F \tag{11.72}$$

$$Q_{kl} = 0 \quad k < F \text{ and } l < F \tag{11.73}$$

we can obtain a simple estimate of the effective potential. Inserting these expressions in (11.70) we have

$$V_{eff}^{elho} = 2\mathrm{Re} \sum_{k=1}^{F} \sum_{l=F+1}^{N} H_{kl} \frac{\cos\left(\sqrt{A_{kk}^{(2)}}\right) \sin\left(\sqrt{A_{kk}^{(2)}}\right)}{\sqrt{A_{kk}^{(2)}}} \times (-i) A_{kl}^{(1)*} \exp\left(\frac{i}{\hbar}(\varepsilon_k - \varepsilon_l)t\right) \tag{11.74}$$

which may be written as

$$V_{eff}^{elho} = 2\mathrm{Re} \sum_{k=1}^{F} \sum_{l=F+1}^{N} j_0\left(2\sqrt{A_{kk}^{(2)}}\right) \frac{H_{kl}}{\hbar} \int_{t_0}^{t} dt' H_{kl}^{*} \sin(\omega_{kl}(t - t')) \tag{11.75}$$

where we have used the fact that

$$j_0(2x) = \sin(2x)/2x \tag{11.76}$$

and introduced $\omega_{kl} = (\varepsilon_k - \varepsilon_l)/\hbar$. Thus the effective potential responsible for electronic "friction" is rather simple in this model. The potential is added to the similar potential for friction due to phonon excitation (see Chapter 8) and coupled self-consistently to the integration of the trajectory for the incoming particle. The effective potential arising due to electron-hole pair excitation, and derived above,

TABLE 11.2 Energy Transfer to Phonons E_{int} in a Cu(111) Crystal by Collisions with Hydrogen and Deuterium

Molecule	E_{int} kJ/mol	
H_2 and D_2	0.0	No substrate coupling
H_2	9.0	Coupling to phonons
D_2	18.9	Coupling to phonons
H_2	9.5	Coupling to phonons and electrons
D_2	20.2	Coupling to phonons and electrons

The surface temperature is 300 K, incident kinetic energy 55.8 kJ/mol, and angles $(\theta, \phi) = (0, 0)$. The molecule is initially in its ground vibrational/rotational state.

has been used to investigate the relative importance of phonon and electronic friction (see Fig. 11.4 and Table 11.2). The electronic friction is very sensitive to the screening parameters and of course also to the charge transfer parameter δ. The value of the charge transfer depends on the distance from the surface and also on the site and molecule bond length. δ can be obtained by Mulliken charge

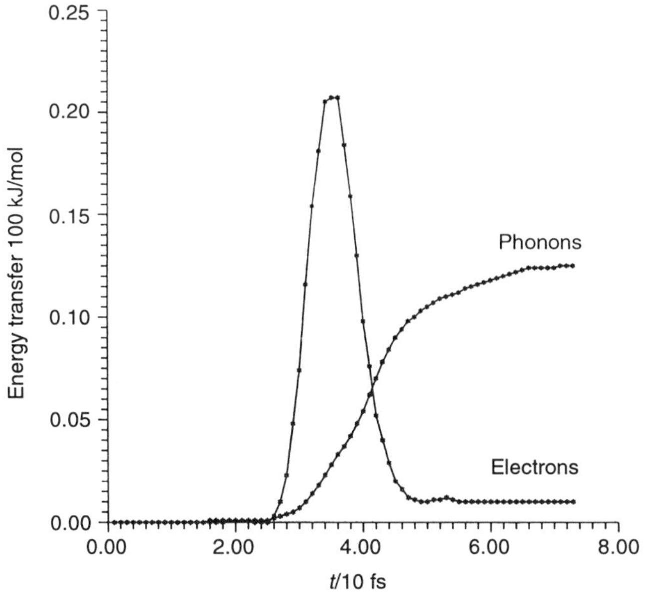

FIGURE 11.4 Electronic and phonon excitation as a function of time for a hydrogen molecule colliding with a copper surface. The initial kinetic energy is 55.8 kJ/mol, incident angles $\theta, \phi = 0.0$, vibrational/rotational state $(v, j) = (0, 0)$, surface temperature 300 K. The screening length in the Fermi-Thomas expression is set to $a = 0.68$ Å^{-1}. The final energy transfer to the phonons is 12.5 kJ/mol and to electrons 1.0 kJ/mol. The other panel shows the distance from the center of mass to the surface $R(t)$ and the bond length $r(t)$ as a function of time. Time unit is 10^{-14} sec = 10 fs.

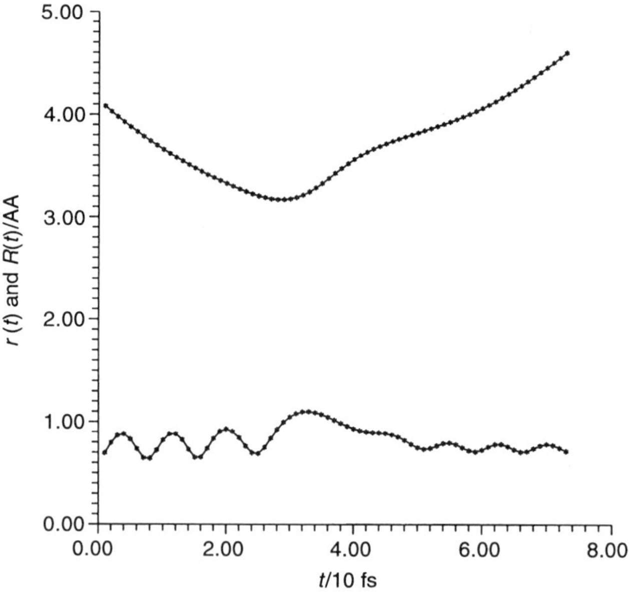

FIGURE 11.4 *(Continued)*

analyses, and typical values are believed to be around 5 to 20% of an electron. If the Fermi-Thomas model is used then the calculations on hydrogen-copper show that the phonon mechanism dominates. The effect of electronic excitation is then primarily that it increases the residence lifetime at the surface, and hence also that the coupling to the phonons is increased (see Table 11.2). Figure 11.4 shows that the coupling to the substrate electrons is strong near the turning point of the trajectory.

The model used here for electronic friction can be improved by using Kohn-Sham one-particle wavefunctions instead of those based on the Bloch potential. Also the solution of the many-electron problem can be improved. Instead of using the sudden limit, we may, with present computer facilities, integrate the eqs. (11.43) for a system involving 100 to 1000 electrons.

In order to evaluate the "hopping integral" H_{kl} we have assumed a fairly simple model, in which the excess charge on the atom/molecule excites the electrons. Other models for evaluation of the non-adiabatic coupling terms (see eq. (11.6)) could be introduced instead and used when evaluating the effective potential (11.75).

11.4 PHONON AND ELECTRON COUPLING

We have seen that it is possible to derive frictional forces arising from coupling of the molecule motion to inelastic excitations in the solid. Thus with the models developed in Chapter 8 and in this chapter we find that the motion of the gas-phase atoms is governed by an effective potential

$$V_{eff} = V_0 + V_{add}(t, T_s) + V_{el}(t) \tag{11.77}$$

where the first term is the intermolecular interaction between the molecule and the solid with the solid atoms placed in their equilibrium positions. The next term is a dynamical potential arising from the coupling to the phonons. It depends upon time and the surface temperature and is defined by a sum over the phonon modes

$$V_{add} = \sum_k \eta_k(t, T_s) V_k^{(1)} \tag{11.78}$$

where $\eta_k(t, T_s)$ is defined by eq. (8.116). The last term is the effective potential arising from the coupling to electron-hole pair excitation $V_{el} = V_{eff}^{elho}$ defined by eq. (11.75). In Figures 11.5 and 11.6 we have shown the terms entering the effective potential for deuterium colliding with a Cu(001) surface consisting of 130 atoms placed in 4 layers. The initial kinetic energy is 77.5 kJ/mol, and incident angles $\theta = \phi = 0°$. Figure 11.5 shows a reflected trajectory where the molecule hits a spot at the surface where V_0 is positive (repulsive). The extra potential arising from the phonon coupling V_{add} is similar in magnitude to the

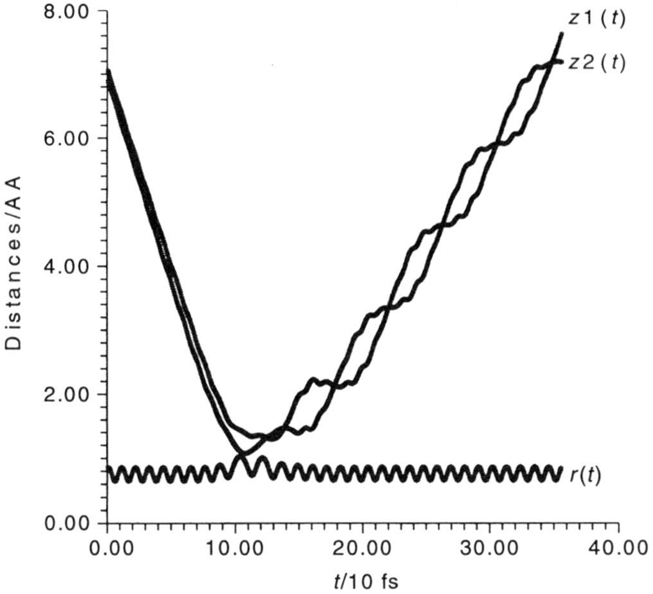

FIGURE 11.5 The upper panel shows the intermolecular bond distance $r(t)$ and the distances $z_i(t)$ $(i = 1, 2)$ from the surface for the two deuterium atoms in D_2 colliding with a copper surface at 300 K. The initial kinetic energy is 77.5 kJ/mol and initial vibrational/rotational state $v = j = 0$. The lower panel shows the static potential V_0, the potential V_{add} arising from the phonon coupling, and the energy transfer to the solid E_{int} as a function of time (in units of 10 fs).

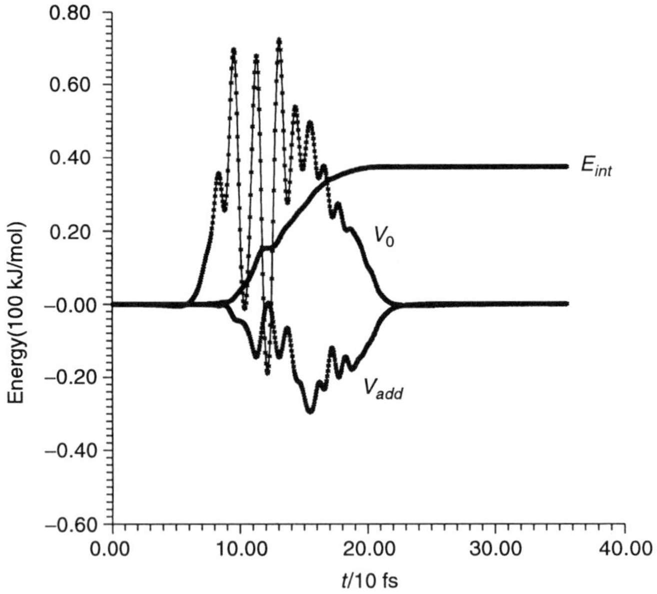

FIGURE 11.5 *(Continued)*

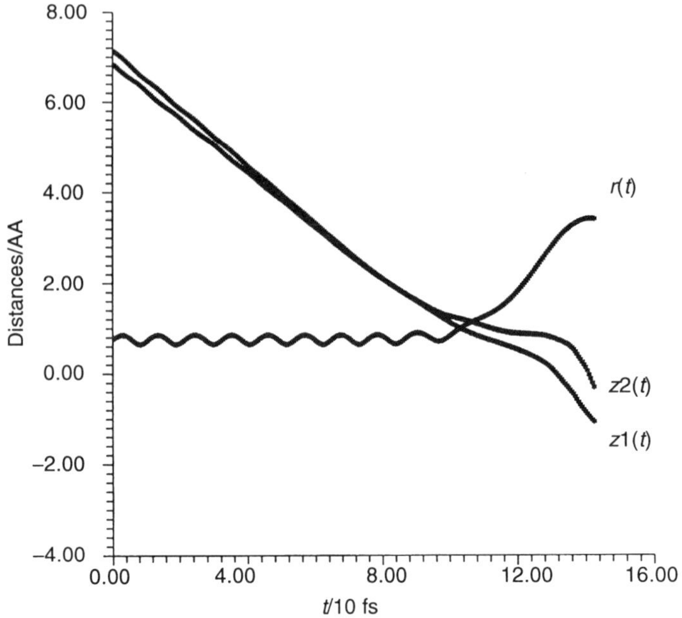

FIGURE 11.6 The upper panel shows the intermolecular bond distances $r(t)$ and $z_i(t)$ $(i = 1, 2)$ for a dissociative trajectory. The lower panel shows the static potential V_0, the potential arising from the phonon coupling V_{add}, and coupling to the surface electrons V_{el}.

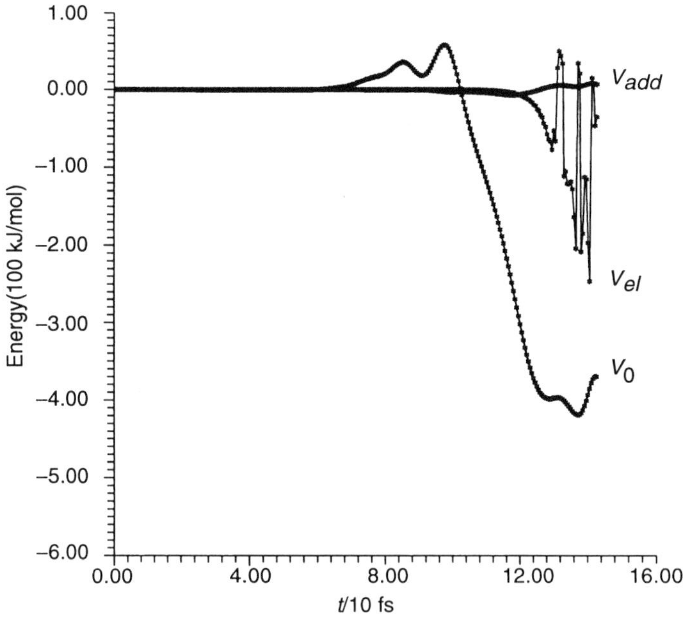

FIGURE 11.6 *(Continued)*

static potential V_0. The potential V_{el} is small and has no influence on the collision. The energy transfer to the solid is E_{int}. It amounts to about 37.6 kJ/mol, i.e., 49% of the initial kinetic energy. If, on the other hand, the molecule comes closer than about 1 Å from the surface, the coupling to the electrons starts to dominate. This is the case for the dissociative trajectory shown in Figure 11.6. In this case the contribution to the effective potential from V_{el} is nonnegligible. These results show that substrate coupling, both the electron and the phonon coupling, do play an important role in molecule surface scattering. Electron-hole pair excitation appears to be most important for dissociative trajectories where the atoms come close to the surface and may even go subsurface.

EXERCISES

11.1: Show that the nonadiabatic coupling term $\langle \phi_n \nabla \phi_m \rangle$ in eq. (11.6) for $n \neq m$ can be written as

$$\frac{\langle \phi_n | \nabla H_{el} | \phi_m \rangle}{E_m - E_n} \tag{11.79}$$

11.2: Show eq. (11.38). *Hint*: Use the fact that the number of states on the energy shell is $(d/dE)N(E)$, where $N(E)$ is the number of states with energy less than E. Also use the relation $\int \sin^2 ax\, dx = (1/2)x - (1/4a)\sin 2ax$.

11.3: Show that the phase factor in eq. (11.17) makes the wavefunction antisymmetric in integralchange of two particles.

11.4: Determine the commutator $[O_{ij}, O_{kl}]$ where $O_{ij} = b_i^+ b_j$ and $O_{ij} = c_i^+ c_j$, i.e., expressed by boson or fermion operators.

11.5: Use the result from exercise 11.3 to show that

$$U_0^+ H_{ij} O_{ij} U_0 = B_{ij} O_{ij} \tag{11.80}$$

where B_{ij} is given by eq. (11.44).

12

PERSPECTIVE

In the present text we have attempted to present the theories needed for a molecular dynamics approach to surface processes. The theories are somewhat more complex and sophisticated than those known from gas-phase dynamics. It is necessary to incorporate the electron and nuclear dynamics of the solid for a complete description of the phenomena. Although this has been known for at least two decades, development in that direction has been going rather slowly. Our approach has still largely been built on the adiabatic principle or the Born-Oppenheimer separation of the electronic and nuclear dynamics. The combination of density functional theory with the use of pseudopotentials for modeling electron-ion interaction has made it possible to solve the electronic structure problem for the ground state energy for systems with up to about 100 or more atoms, depending on the sophistication of the exchange-correlation term. Thus the progress, as far as the electronic part of the problem is concerned, has been extensive and significant. The progress in DFT theory, as well as traditional ab initio calculations, was therefore recognized with the Nobel prize awarded to Pople and Kohn in 1998. With increasing development of computer power and methodology the number of electrons which can be handled will quickly be pushed to much higher values. However, the solution of the many-body problem requires a potential of full dimension in the nuclear coordinates. This is, and will in the foreseeable future, be possible only by using some parametric representation of the surface. The combination of the diatomics of molecules (DIM) method known from gas-phase dynamics and density functional theory appears to be a useful method. Alternatively the introduction of direct dynamics methods, in which the electronic structure calculation is performed on the fly, is a possible avenue. Recently a trajectory-based second quantization method for the nuclear dynamics has been proposed [268]. This method makes it possible to solve the nuclear dynamics problem within a quantum mechanical framework rather than classically, as is proposed in the Carr-Parinello method (see section 10.3).

Since surface reconstruction under chemisorption processes is taking place [269], it is an open question how this is best accounted for. Should the atoms be allowed to relax adiabatically on an, in principle, infinite timescale, or should one use the timescale on which the collision occurs? Is the Born-Oppenheimer separation meaningful under such circumstances? The interplay between nuclear and electronic degrees of freedom is not fully understood in these cases.

Molecule-surface scattering calculations have been carried out treating all six dimensions of the diatomic (hydrogen) quantum mechanically [270, 271]. Comparison [270, 272] with calculations using a classical mechanical treatment of the dynamics confirmed results known from gas-phase dynamics, namely that average quantitites often are well predicted using classical dynamics, whereas state-resolved quantities are often poorly predicted by classical mechanics (for an early study see ref. [273]).

However, the coupling to the substrate has been neglected in these calculations. Although this coupling may be neglected for light molecules such as hydrogen, such calculations are only of academic interest for heavier molecules. The calculations showed, as mentioned, that classical treatment of the dynamics was accurate for predicting the sticking probability. However, the situation is complicated by the fact that if the substrate coupling is included then the classical treatment of the vibrational motion of the diatom is hampered by the availability of the zero-point energy of the molecule, which when available can be used to surpass the reaction barrier [97]. If more complicated problems such as the influence of surface structure on catalytic reactivity are considered (see, e.g., [274, 275]), it is necessary to treat the full dynamic problem in a realistic manner. This treatment must also include the substrate dynamics.

Although the present text has not been devoted to kinetic modeling of surface reactions, it is obvious that many important industrial processes are at present only understood at the kinetic level. Hence the discussion of the mechanism which at best describes the measured kinetics is an important issue (see, e.g., [276] for modeling of the NO + CO kinetics on Rhodium). However, it is well known that different mechanisms may give the same or very similar reaction rate equations. The fact that the principle of detailed balance, which is strictly applicable only for elementary reactions with assumed thermal equilibrium distributions of reactants and intermediates, needs to be applied in order to obtain reverse rates from the forward rates and the equilibrium constant also adds to the complexity of kinetic modeling. This is especially so in cases where the assumption of an equilibrium distribution on the molecular energy levels is questionable. Molecules do have degrees of freedom, especially the vibrational degrees of freedom, which equilibrate on timescales longer than those of many chemical reactions. Hence the kinetic modeling should take this aspect into account in order to be reliable.

New experimental as well as theoretical tools, which take the molecular point of view, are necessary to resolve some of the controversies concerning molecular mechanisms. However, kinetic modeling will still be an important part of the theoretical tool box for making contact with experiments carried out under bulk conditions, or for predicting improvements of existing catalytic reactors. Such modeling is carried out by integrating the rate equations according to the proposed mechanism, or by using a Monte Carlo approach [277].

An interesting development and possibility is the study of reactivity of clusters, surface anomalities as steps, or the size effect of islands on catalytic activity [278]. To study such processes by using some of the methods presented here is not more complicated than to study regular surfaces. In principle, therefore, the design of a catalyst from a solution of equations or models with roots in the fundamental quantum mechanical equations is within reach [279].

It is important to realize that the problem concerning chemical reactivity at surfaces is a problem which concerns not only the structure and energetics at the surface but also the proper sampling over and incorporation of the various degrees of freedom in which the available energy may dissipate. Thus, internal degrees of freedom of the involved molecules, electronic states of the molecules as well as the substrate, and the vibrational motions of the surface atoms all need to be considered. The theoretical tools for achieving this goal are in principle available. Some of them need further investigation and refinements — a process which will take place in the coming years. This process will lead to the ultimate goal of understanding catalytic processes from first principles.

APPENDIX A
RATE CONSTANTS

Rate constants for chemical reactions can be obtained theoretically by various methods. We shall only mention two here, namely the collision theoretical approach and the approximate "transition state" theory (TST). In the collision theory the reaction rate constant is obtained by integrating over a Boltzmann-weighted flux of particles hitting a target. In gas-phase dynamics, the target is another molecule in the gas-phase; in surface dynamics it is either the surface itself (sticking, dissociation) or an adsorbed molecule. In any case, the rate constant is obtained as

$$k(T) = \int dE_{kin} f(E_{kin}, T) v \sigma_{target} P_{event} \qquad \text{(A.1)}$$

where $f(E_{kin}, T)$ is a distribution function of the kinetic energy in the relative center of mass motion of the projectile and the target molecule. For surface reactions the relative motion is often (unless a diffusing or moving target at the surface is considered) just the center of mass motion of the incoming molecule (projectile). For bulk situations $f(E_{kin}, T)$ is the Boltzmann distribution, but it may also be a narrow distribution function as provided in beam experiments. The relative velocity is denoted by v, and the size of the target, the cross section, is σ_{target}. Finally, P_{event} is the probability for the event of interest (reaction, sticking, dissociation, etc.) to occur. The unit of the rate constant is $\text{cm}^3\text{sec}^{-1}\text{molecule}^{-1}$ and the reaction rate for a gas-phase reaction $A + B \rightarrow C + D$ is obtained as

$$r = k(T)[A][B] \qquad \text{(A.2)}$$

where $[A]$ denotes the concentration in molecules/cm^3. Thus the unit of r is molecule $\times$ $\text{sec}^{-1}\text{cm}^{-3}$. For surface reactions the rate is

$$r = k(T)[A]\rho_B \qquad \text{(A.3)}$$

where ρ_B is the density of B molecules at the surface in molecules/cm^2. Hence the unit of the reaction rate is molecule $\times$ $\text{sec}^{-1}\text{cm}^{-2}$, i.e., number of molecules reacting per unit time and unit area. In classical trajectory calculations, the cross section, or size, of the target is taken as the "aiming area" A, the size of a unit cell area. The target molecule is placed at the surface and allowed to relax or move according to the situation one wants to model. The initial condition of the projectile is then taken (see Chapter 8) such that a random point in the aiming area is hit (if there is no interaction or corrugation). The event of the collision determines P_{event}.

Many surface reactions have an activation barrier and are therefore likely to go through an activation complex when reacting. Here we should keep in mind that there may be many different barriers (depending upon the surface site) and that before using the result of transition state theory one has then to consider the probability for reaching a particular activated complex at a given site. Here we simply incorporate a steric or statistical factor (P_{steric}) in order to indicate that this effect should not be forgotten.

For recombination reactions through the Eley-Rideal (E-R) mechanism we can encounter situations with little or no activation barrier. Molecular dynamics studies of recombination at the surface through the E-R mechanism [282] show that the steric factor P_{steric} may be small, due to dynamic factors. However, if we assume an activation barrier, we can apply the transition state theory. According to transition state theory the reaction rate is obtained as

$$r = \nu[AB]^* = k[AB][*] \tag{A.4}$$

where ν is the attempt frequency, i.e., the number of times the transition state complex attemps to cross the barrier in each time unit. $[AB]^*$ is the transition state concentration, in molecules per unit area. $[AB]$ is the gas-phase concentration in molecules per cubic centimeter, and $[*]$ the concentration of transition state sites per unit area, proportional to A^{-1}. Hence the unit of the rate is $\sec^{-1}\text{cm}^{-2}$. In transition state theory the attempt frequency is evaluated to be

$$\nu = \frac{k_B T}{h} \tag{A.5}$$

If the reaction occurs through an Eley-Rideal type of process, i.e., from the gas-phase to the transition-state, then we have, according to the classical transition state theory, an assumption of equilibrium between the gas-phase and transition state concentrations, that is,

$$[AB]^* = K[AB] \tag{A.6}$$

where $[AB]$ is the gas-phase concentration. According to statistical mechanics, we have, for an equilibrium situation [130], that

$$\Pi_i N_i^{\nu_i} = \Pi_i Q_i^{\nu_i} \exp(-E_0/k_b T) \tag{A.7}$$

where N_i is the number of molecules i, ν_i the stoichiometric coefficient, Q_i the partition function for molecules i, and E_0 the energy difference (in electronic energy) between reactant and transition state molecules. Thus, for the specific case considered, we have

$$\frac{N^*_{AB}}{N_{AB}} = \frac{Q^*_{AB}}{Q_{AB}} \exp(-E_0/k_B T) \qquad (A.8)$$

Thus our rate can be written as

$$r = P_{steric} \nu A^{-1} \frac{Q^*_{AB}}{Q_{AB}} V[AB] \exp(-E_0/k_B T) \qquad (A.9)$$

where we have used the fact that $[AB^*] = N^*_{AB}/A$ and $[AB] = N_{AB}/V$. Hence the rate constant can be defined as

$$k(T) = P_{steric} \frac{k_B T}{h} \frac{Q^*}{Q_I} \exp(-E_0/(k_B T)) \qquad (A.10)$$

where Q^* and Q_I are the partition functions for the activated complex and the reactants respectively. If the electronic state of the reactants or the transition state is degenerate, this needs to be incorporated by an additional electronic degeneracy factor g_{el}. E_0 is the barrier for reaction, and T the temperature. If the atom B is adsorbed and the atom A comes from the gas phase, the above expression is modified to

$$r = P_{steric} \nu \frac{V Q^*_{AB}}{Q_A Q_B} [A][B] \exp(-E_0/(k_B T)) \qquad (A.11)$$

where $[A]$, the gas-phase concentration, can be expressed by the gas pressure P_A as P_A/RT. The rate constant for this situation is

$$k(T) = P_{steric} \frac{k_B T}{h} \frac{V Q^*_{AB}}{Q_A Q_B} \exp(-E_0/(k_B T)) \qquad (A.12)$$

For an atom A we have only the translational partition function to evaluate, i.e., $Q_A = V Q_t$, where Q_t is given below. The surface-bound atom B has three vibrational degrees of freedom, and hence the partition function is

$$Q_B = \Pi^3_{i=1} \frac{1}{2 \sinh(\hbar\omega^*_i/2k_B T)} \qquad (A.13)$$

where ω^*_i are the vibrational frequencies of the surface-bound atom.

The activation energy is

$$E_0 = E^*_{AB} - E_A - E^*_B \qquad (A.14)$$

where the energies refer to the electronic energies of the transition state E^*_{AB}, the surface-bound atom E^*_B, and the gas-phase atom E_A.

If we consider a diatomic molecule instead of a gas-phase atom, we obtain the following partition function Q_I

$$Q_I = Q_t Q_r Q_v \tag{A.15}$$

where Q_t is the translational partition function per unit volume

$$Q_t = \frac{1}{h^3}(2\pi m k_B T)^{3/2} \tag{A.16}$$

Here m is the mass of the molecule. The rotational and vibrational partition functions for a diatomic molecule are [130]

$$Q_r = \frac{2Ik_BT}{\sigma\hbar^2} \tag{A.17}$$

$$Q_v = \frac{1}{2\sinh(0.5\hbar\omega/k_BT)} \tag{A.18}$$

where I is the moment of inertia, σ a symmetry factor equal to 2 for homonuclear and 1 for heteronuclear diatomic molecules. The vibrational frequency is ω. The molecule at the transition state has 6 vibrational frequencies, one of which is imaginary. The imaginary frequency represents the degree of freedom connected to the motion over the transition state. It does not enter the classical transition state theory, but can be used to estimate a tunneling correction factor. Thus the partition function for the transition state is

$$Q^* = \Pi_{i=1}^{5} \frac{1}{2\sinh(0.5\hbar\omega_i^*/k_BT)} \tag{A.19}$$

In more modern transition state theory, the symmetry factors are omitted in the rotational partition functions and combined in a common symmetry factor in the expression for the rate constant. This symmetry factor counts the number of equivalent ways of forming the transition state (see, e.g., [130, 280]).

In the case of a Langmuir-Hinshelwood (L-H) type of the reaction, where adsorbed atoms A and B react, we have the following reaction rate per unit time and surface area

$$r = k[A][B] \tag{A.20}$$

that is, the unit of the reaction rate is the same as before. But since the two concentrations are given in molecules/cm^2, the unit of the reaction rate constant becomes sec^{-1}cm^2. The equilibrium assumption of the transition state theory now gives

$$\frac{N_{AB}^*}{N_A N_B} = \frac{Q_{AB}^*}{Q_A Q_B}\exp(-E_0/k_BT) \tag{A.21}$$

where E_0 is the energy difference between the transition state and the energy of the two adsorbed atoms, A and B. The temperature T is now the surface temperature. Again we have the reaction rate defined as the rate with which the transition state is passed, i.e., as

$$r = \nu[AB]^* = \nu A \frac{Q^*_{AB}}{Q_A Q_B} \exp(-E_0/k_b T)[A][B] \qquad \text{(A.22)}$$

where we have introduced $[A] = N_A/A$, and so on. The rate constant in this case is defined by

$$k(T) = \frac{k_b T}{h} \frac{Q^*_{AB}}{Q_A Q_B} A \exp(-E_0/k_b T) \qquad \text{(A.23)}$$

where the partition functions all are for surface-bound atoms or molecules, i.e., AB^* has five vibrational and the atoms three vibrational degrees of freedom. The activation energy is

$$E_0 = E^*_{AB} - E^*_A - E^*_B \qquad \text{(A.24)}$$

where the * indicates that the species is surface bound. The difference in activation barrier between the L-H and the E-R situation is then $E_A - E^*_A$, which is the binding energy of the atom A. That is, if the atom is bound strongly to the surface, the E-R mechanism may dominate over the L-H mechanism.

We notice that we have incorporated the vibrational zero-point energy in the partition functions. Sometimes it is left out of the partition function and included in the activation energy instead, such that

$$\Delta E_0 = E_0 + \frac{\hbar}{2} \sum_i (\omega^*_i - \omega_i) \qquad \text{(A.25)}$$

and hence we obtain the vibrational partition function as $Q_i = 1/(1 - \exp(-\hbar\omega_i/k_B T))$, which in the high temperature limit becomes $Q_i \sim k_B T/\hbar\omega_i$.

Improvement of the conventional TST theory can be obtained by the canonical variational transition state theory in which the location of the transition state is searched for by looking for the position along the reaction path where the reactive flux is minimum [283]. Alternatively a dynamical correction to the transition state theory can be calculated by numerical evaluation of the transmission factor [31]. The transmission factor should include the effect of recrossing over the transition state, i.e., a factor which is smaller than unity. The tunneling correction factor increases the reaction rate and depends strongly upon the mass of the atom or molecule, the barrier height, and the barrier width. In surface scattering, we have, in principle, an additional correction factor which has been denoted a steric factor. This factor should account for the tendency of reaching the most favorable reaction pathway. This steric factor will depend on kinetic energy, surface temperature, and approach angle. It can be estimated only by running trajectories over the surface. This factor makes the use of transition state theory for surface reactions somewhat more problematic than in gas-phase chemistry.

The TST is a specific limit of a more general theory called the reaction path theory [156]. The advantage of reaction path theory over TST is that vibrationally resolved rate constants can be obtained from the theory. The information needed in order to define the reaction path and the hamiltonian which governs the dynamics are gradients and hessians (second derivative matrices) along the reaction path. The gradients are used to map out the path, and the hessians to derive the coupling terms for coupling of the motion along the path to the vibrational motion perpendicular to it. Since this information is exactly what is available in ab initio program packages the reaction path approach makes it possible to couple electronic structure calculations and dynamics in a convenient manner.

APPENDIX B

VALENCE BOND TREATMENT OF FOUR-ELECTRON SYSTEMS

We introduce a set of one-electron wavefunctions ϕ_i, consisting of n orthonormal wavefunctions. For the binding orbitals we assume paired electrons; and with an even number of electrons, we would look for functions which are eigenfunctions to the operator S_z with the eigenvalue 0, where S_z

$$S_z = \sum_{i=1} s_z^i \tag{B.1}$$

is a sum over the individual electron spin operators. If we consider just 4 electrons, the one-electron wavefunctions can be denoted a, b, c, and d for simplicity. With the requirement that S_z working on the wavefunction should give zero, we need the number of α and β spin to be the same, i.e., $n_\alpha = n_\beta = n/2$. For n electrons we can form

$$\binom{n}{\frac{n}{2}} \tag{B.2}$$

of such functions. Thus with $n = 4$ we have 6 possible functions

$$\phi_1 = |a\bar{b}c\bar{d}| \tag{B.3}$$

$$\phi_2 = |\bar{a}b\bar{c}d| \tag{B.4}$$

$$\phi_3 = |\bar{a}\bar{b}cd| \tag{B.5}$$

$$\phi_4 = |\bar{a}bc\bar{d}| \tag{B.6}$$

$$\phi_5 = |a\bar{b}\bar{c}d| \tag{B.7}$$

$$\phi_6 = |ab\bar{c}\bar{d}| \tag{B.8}$$

where $\overline{a}$ means that the electron in orbital a has the spin α. The spin function of the stable bond should be antisymmetric in the interchange of two electrons. If we require a bond between a and b and between c and d, we have

$$\psi_1 = (\phi_1 - \phi_4 - \phi_5 + \phi_2) \tag{B.9}$$

constructed by requiring opposite spin on the two sites. Likewise for a bond between a and d and between b and c

$$\psi_2 = (\phi_6 - \phi_1 - \phi_2 + \phi_3) \tag{B.10}$$

Next consider the secular equation for the problem

$$\begin{vmatrix} H_{11} - ES_{11} & H_{12} - ES_{12} \\ H_{21} - ES_{21} & H_{22} - ES_{22} \end{vmatrix} = 0 \tag{B.11}$$

where $H_{11} = \langle\psi_1|H|\psi_1\rangle$, $S_{11} = \langle\psi_1|\psi_1\rangle$ etc. Considering first the matrix elements with the primitive functions ϕ_i, we have [174]

$$\begin{aligned} h_{12} = \langle\phi_1|H|\phi_2\rangle = 0 \; h_{13} = -J_{ad} \; h_{14} = -J_{ab} \\ h_{15} = -J_{cd} \; h_{16} = -J_{bc} \; h_{23} = -J_{bc} \; h_{24} = -J_{cd} \\ h_{25} = -J_{ab} \; h_{26} = -J_{ad} \; h_{34} = -J_{bd} \; h_{35} = -J_{ac} \\ h_{46} = -J_{ac} \; h_{36} = 0 \; h_{45} = 0 \; h_{56} = -J_{bd} \end{aligned} \tag{B.12}$$

where J_{xy} are exchange integrals. For the diagonal elements we obtain

$$\begin{aligned} h_{11} = Q - J_{ac} - J_{bd} \; h_{22} = Q - J_{ac} - J_{bd} \\ h_{33} = Q - J_{ab} - J_{cd} \; h_{44} = Q - J_{ad} - J_{bc} \\ h_{55} = Q - J_{bc} - J_{ad} \; h_{66} = Q - J_{ab} - J_{cd} \end{aligned}$$

where Q is a sum of Coulomb integrals

$$Q = Q_{ab} + Q_{cd} + Q_{ac} + Q_{ad} + Q_{bc} + Q_{bd} \tag{B.13}$$

Using these expressions in the secular equation we obtain

$$H_{11} = 4Q - 2(J_{ac} + J_{ad} + J_{bc} + J_{bd}) + 4J_{ab} + 4J_{cd} \tag{B.14}$$

$$H_{22} = 4Q - 2(J_{ab} + J_{ac} + J_{cd} + J_{bd}) + 4J_{ad} + 4J_{bc} \tag{B.15}$$

$$H_{12} = -2Q - 2(J_{ab} + J_{ad} + J_{bc} + J_{cd}) + 4(J_{ac} + J_{bd}) \tag{B.16}$$

We furthermore have $S_{11} = S_{22} = 4$ and $S_{12} = -2$. The energy of the lowest state is finally obtained as

$$E = Q - \left\{\tfrac{1}{2}[(J_{ab} + J_{cd} - J_{ad} - J_{bc})^2 + (J_{ad} + J_{bc} - J_{ac} - J_{bd})^2 + (J_{ac} + J_{bd} - J_{ab} - J_{cd})^2]\right\}^{1/2} \tag{B.17}$$

This expression obtained by London [243] for the four-electron system (see Fig. 4.3) can also be used for a three-electron system by removing all integrals containing the index d. This gives the LEPS (London, Eyring, Polanyi and Sato)-surface discussed in Chapter 4.

APPENDIX C

HARTREE AND HARTREE-FOCK METHODS

The treatment of the many-body system of nuclei and electrons of which solids and metals consist, makes it necessary to introduce approximations. Traditional approximations in electronic structure theory are those due to Hartree and Fock. We shall here only briefly give the main equations relevant for the discussion of the free-electron gas results (Chapter 9) and the density functional theory (Chapter 10). Consider, for instance, a metal of N atoms and each atom having Z electrons, where the number of atoms is of the order Avogadros number. The number of electrons to be considered for each atom can be lowered from the actual number to the number of valence electrons by introducing effective core potentials. The nonrelativistic Schrödinger equation for the electronic part of the problem is then

$$H = -\sum_{i=1}^{NZ} \frac{\hbar^2}{2m} \nabla_i^2 - \sum_{i=1}^{NZ} \sum_{a=1}^{N} V_{ia}(R_{ai}) + \frac{1}{2} \sum_{i \neq j}^{NZ} \frac{e^2}{r_{ij}} + \frac{1}{2} \sum_{a \neq b} \frac{Z^2 e^2}{R_{ab}} \tag{C.1}$$

where $V_{ia}(R_{ai})$ is the effective core potential between electron i and ion core a. If core potentials are not introduced, this term is the usual Coulomb attraction

$$V_{ia}(R_{ai}) = \frac{Ze^2}{R_{ai}} \tag{C.2}$$

R_{ai} is the distance between the center of core and electron i, r_{ij} the distance between the two electrons i and j, and R_{ab} the distance between the two core centers a and b. Here we have assumed that the two ion cores are sufficiently far apart for the interaction to be a simple Coulomb repulsion. The first term in the hamiltonian is the kinetic energy of the electrons and the third term the electronic repulsion. The ground state or lowest energy is then obtained by solving the time-independent Schrödinger equation with the above operator. This is, however, not a trivial task, and considerable simplification is therefore needed. One such simplification is obtained if it is assumed that the wavefunction can be expressed as a simple Hartree product

$$\Psi = \psi_1(\mathbf{x}_1)\psi_2(\mathbf{x}_2)\ldots\psi_N(\mathbf{x}_N) \tag{C.3}$$

where the notation $\mathbf{x}_i$ denotes space and spin coordinates. Also one may assume that the one-electron functions are orthorgonal, i.e.,

$$\int d\mathbf{x}\psi_i^*(\mathbf{x})\psi_j(\mathbf{x}) = \delta_{ij} \tag{C.4}$$

The functions can furthermore be approximated by a simple product form of the space ϕ and spin functions ξ, i.e.,

$$\psi_i(\mathbf{x}) = \phi_i(\mathbf{r})\xi_i \tag{C.5}$$

where ξ_i is either an α or a β spin function. Inserting now the product wavefunction in the expression

$$E_0 = \int d\tau\Psi^* H\Psi \tag{C.6}$$

using the orthogonality relations of both spin and spatial functions, one obtains

$$E_0^{hartree} = \sum_i \int \psi_i^*(\mathbf{r})\left[-\frac{\hbar^2}{2m}\nabla^2 + \sum_{a=1}^{N} V_a(\mathbf{r})\right]\psi_i(\mathbf{r})d\mathbf{r}$$
$$+\frac{1}{2}\sum_{i\neq j}^{N}\iint \frac{e^2}{r_{ij}}|\phi_i(\mathbf{r}_i)|^2|\phi_j(\mathbf{r}_j)|^2 d\mathbf{r}_i d\mathbf{r}_j + \frac{1}{2}\sum_{a\neq b}\frac{e^2}{R_{ab}} \tag{C.7}$$

where, for simplicity, we, have set $Z = 1$. This expression for the energy is not expected to be very accurate, since the trial wavefunction has not been properly antisymmetrized. Thus the wavefunction must be antisymmetric for interchange of two electrons (fermions). This can be introduced by constructing a determinantal wavefunction of the form

$$\Psi = \frac{1}{\sqrt{N!}}|\psi_1(\mathbf{x}_1)\ldots\psi_N(\mathbf{x}_N)| \tag{C.8}$$

Inserting this expression in eq. (C.6) gives after some manipulations (see, e.g., refs. [248]) the following expression for the energy

$$E_0 = E_0^{hartree} + E_{ex} \tag{C.9}$$

that is, the same expression as before except for an additional "exchange" term

$$E_{ex} = -\frac{1}{2}\sum_{ij}\iint \frac{e^2}{r_{ij}}\phi_i^*(\mathbf{r}_i)\phi_j^*(\mathbf{r}_j)\psi_i(\mathbf{r}_j)\psi_j(\mathbf{r}_i)d\mathbf{r}_i d\mathbf{r}_j \tag{C.10}$$

where the sum is implicitly understood to be over "parallel" spins or the same spin function on both orbitals. Again the expression is given for $Z = 1$ but with rather obvious extensions to the general case. From the Hartree and Hartree-Fock theories we can also obtain, by applying the variational principle, an equation (Fock equation) for the one-electron wave functions ϕ_i. Thus the Fock equation is

$$\left[-\frac{\hbar^2}{2m}\nabla_i^2 + \sum_{a=1}^{N} V_a(\mathbf{r_i}) + e^2 \sum_{j=1}^{N} \int \frac{|\phi_j(\mathbf{r}_j)|^2}{r_{ij}} d\mathbf{r}_j\right] \phi_i(\mathbf{r}_i)$$
$$- e^2 \sum_{j=1}^{N} \left[\int \frac{\phi_j^*(\mathbf{r}_j)\phi_i(\mathbf{r}_j)\phi_j(\mathbf{r}_i)}{r_{ij}\phi_i(\mathbf{r}_i)}\right] \phi_i(\mathbf{r}_i) = \varepsilon_i \phi_i(\mathbf{r}_i) \qquad \text{(C.11)}$$

where the operator in the first bracket is the one obtained from the Hartree theory and the second the correction due to exchange. We note that the first term is the kinetic energy term of electron i and the second two terms constitute an average potential in which the electron moves. The last term is not a proper potential in the same sense as the first but arising from the "antisymmetrization" condition.

The approximation inherent in the Hartree-Fock theory is the lack of proper inclusion of electronic correlation. But the equations are solvable even for many-electron systems. Today the most popular way of treating such systems approximately is through the density functional theory, where the correlation is accounted for by more or less empirical means (see Chapter 10).

APPENDIX D

THE INVERSION PROBLEM

For systems containing a large number of identical particles or excitations, such as fermions and bosons, it is convenient to switch from the "first" quantization description involving the wavefunction to the second quantization involving operators. The operators are creation and annihilation operators, which when working on a state function create or destroy an excitation or a particle. If the dynamics can be expressed in terms of a closed set of operators, then the time-dependent Schrödinger equation can also be algebraically solved in terms of the operators. A set of operators is closed if the operators and commutators between operators in the set also are members of the set. In Chapters 8 and 11 we have considered this situation for bosons and fermions when describing the interaction between the gas-phase atoms and the solid excitations (vibrations and electrons). Thus operators of the type $a_k^+ a_l$ or $c_k^+ c_l$, where a_k and c_k are boson and fermion operators, form closed sets with respect to commutations. In the actual numerical work which has to be carried out, we introduce a function or matrix representation of the operators [61]. The numerical work involves the solution of a matrix equation (for the R-matrix) of the dimension $M \times M$ where M is the number of operators. For a solid consisting of N atoms we have $M = 3N - 6$ boson operators, i.e., a large reduction as compared with a state expansion technique is achieved. The reason is that a boson state can contain an infinite number of excitations (quanta). By contrast, the fermions have restrictions concerning the number of particles, which can be either zero or one. Thus the number of operators and the number of states are comparable. In either case, systems with about 1000 to 2000 atoms or the same number of electrons can be treated with this technique. Furthermore, in Chapter 8 we have formulated a theory where only the boson operators a_k and a_k^+ are included. Then the numerical work is reduced to the calculation of $2 \times (3N - 6)$ integrals.

In Chapters 8 and 11 we considered the solution to the set of equations involved, i.e.,

$$i\hbar\dot{\mathbf{R}}(t, t_0) = \mathbf{B}\mathbf{R}(t, t_0) \tag{D.1}$$

where $\mathbf{R}(t_0, t_0) = \mathbf{I}$ the unit matrix. In order to obtain the amplitude for a transition from an initial state vector $|n_1 \ldots .n_M\rangle$ to $|n_1' \ldots .n_M'\rangle$ (see [61]) we use the evolution operator U_{VV}

$$U_{VV} = \Pi_{kl} \exp(\alpha_{kl}(t, t_0) a_k^+ a_l) \tag{D.2}$$

and hence we need to evaluate the functions α_{ji}. We introduce the new functions

$$f_{ij} = \alpha_{ij} \ (i \neq j) \tag{D.3}$$

$$f_{ii} = \exp(\alpha_{ii}) - 1 \tag{D.4}$$

and obtain the functions f_{ij} in the following manner [61]:

For $m < n$ we have

$$f_{mn} = |\mathbf{A}|/|\mathbf{D}| \tag{D.5}$$

where $|\mathbf{A}|$ indicates the determinant of the matrix $\mathbf{A}$. The matrices $\mathbf{A}$ and $\mathbf{B}$ are of order $(n-1) \times (n-1)$ and the elements of the two matrices given as

$$D_{kl} = Q_{kl}(k \neq l) \tag{D.6}$$

$$D_{kk} = 1 + Q_{kk} \tag{D.7}$$

and

$$A_{km} = Q_{kn} \tag{D.8}$$

$$A_{kl} = D_{kl} \ (l \neq m) \tag{D.9}$$

where k and l runs from 1 to $n-1$. For the element f_{nn} we have

$$f_{nn} = Q_{nn} - \sum_{m=1}^{n-1} f_{mn} Q_{nm} \tag{D.10}$$

where the elements f_{mn} have just been determined. For f_{nm} where $m < n$ we obtain

$$f_{nm} = \frac{Q_{nm} - \sum_{k=1}^{m-1} Q_{nk} f_{km}}{1 + f_{nn}} \tag{D.11}$$

Thus the above equations enable us to determine all the elements f_{nm} from the solution to the differential equation (D.1) using the fact that $\mathbf{Q} = \mathbf{R} - \mathbf{I}$. The expectation value of the hamiltonian involves only the functions Q_{ij}. Thus the above "inversion" problem need not be solved in order to evaluate the effective hamiltonian H_{eff}. This has been utilized when calculating the effective hamiltonians arising from the coupling to phonon and electron excitation in the solid.

RECENT BOOKS AND REVIEW ARTICLES ON MOLECULE SURFACE DYNAMICS

R. B. Gerber, R. Kosloff, and M. Berman, Time-dependent wavepacket calculations of molecular scattering from surfaces, *Comp. Phys. Rep.* **5**, 59–114 (1986).

V. Bortolani and A. C. Levi, Atom-surface scattering theory, *Rivista del Nuovo Cimento* **9**, 1–77 (1986).

R. B. Gerber, Molecular scattering from surfaces: Theoretical methods and results, *Chem. Rev.* **87**, 29–79 (1987).

G. D. Billing, The dynamics of molecule-surface interaction, *Comp. Phys. Rep.* **12**, 383–451 (1990).

A. E. DePristo and A. Kara, Molecule-surface scattering and reaction dynamics, *Adv. Chem. Phys.* **77**, 163–253 (1990).

Physics of Solids and Liquids Interaction of atoms and molecules with solid surfaces, eds. V. Bortolani, N. H. March, and M. P. Tosi, Plenum Press, New York (1990).

G. P. Brivio and T. B. Grimley, Dynamics of adsorption/desorption at solid surfaces, *Surf. Sci. Rep.* **17**, 1–84 (1993).

G. A. Somarjai, *Introduction to Surface Chemistry and Catalysis*, Wiley, New York (1994).

G. R. Darling and S. Holloway, The dissociation of diatomic molecules at surfaces, *Rep. Prog. Phys.* **58**, 1595–1672 (1995).

R. I. Masel, Principles of adsorption and reaction on solid surfaces, Wiley, New York (1996).

B. Hammer and J. K. Nørskov, in *Chemisorption and Reactivity on Supported Clusters and Thin Films*, eds. R. M. Lambert and G. Pacchioni, Kluwer Academic Publishers, The Netherlands, pp. 285–351 (1997).

M. R. Radeke and E. A. Carter, Ab initio dynamics of surface chemistry, *Ann. Rev. Phys. Chem.* **48**, 243–270 (1997).

BIBLIOGRAPHY

[1] H. Brune, *Surf. Sci. Rep.* **31**, 121 (1998).

[2] G. A. Somorjai, *Appl. Surf. Sci.* **121/122**, 1 (1997).

[3] G. Ertl, *Appl. Surf. Sci.* **121/122**, 20 (1997).

[4] I. Langmuir, *Trans. Faraday Soc.* **17**, 607 (1922).

[5] R. Mezaki and H. Inoue, Rate equations of solid catalyzed reactions, University of Tokyo Press, Tokyo (1991).

[6] N. E. Henriksen, F. Y. Hansen, and G. D. Billing, *Chem. Phys. Lett.* **244**, 350 (1995).

[7] S. Brunauer, P. H. Emmett, and E. Teller, *J. Am. Chem. Soc.* **60**, 309 (1938).

[8] M. Cacciatore, M. Rutigliano, and G. D. Billing, *J. Thermophys. and Heat Transfer* **13**, 195 (1999).

[9] J. E. Lennard-Jones, *Trans. Faraday Soc.* **28**, 333 (1932).

[10] W. S. Struve, J. R. Krenos, D. L. McFadden, and D. R. Herschbach, *J. Chem. Phys.* **62**, 404 (1975).

[11] B. Baule, *Ann. Physik* **44**, 145 (1914).

[12] W. H. Weinberg and R. P. Merrill, *J. Vac. Sci. Tech.* **8**, 718 (1971).

[13] K. Christman, O. Scober, G. Ertl, and M. Neuman, *J. Chem. Phys.* **60**, 4528 (1974).

[14] J. K. Nørskov, K. A. Houmøller, P. Johansson, and B. I. Lundquist, *Phys. Rev. Lett.* **46**, 257 (1981).

[15] Named after British mineralogist W. H. Miller (1810–1880).

[16] A. Bravais was a French physicist (1811–1863).

[17] W. H. Bragg, *Phil. Trans. A* **215**, 254 (1915). Bragg experimented with X-ray scattering of KCl and NaCl.

[18] H. Goldstein, *Classical Mechanics*, Addison-Wesley, London (1959).

[19] W. H. Weinberg, in *Dynamics of Gas-Surface Reactions*, eds. C. T. Rettner and M. N. R. Ashfold, Royal Society of Chemistry, Cambridge (1991), pp. 171.

[20] G. A. Somarjai, *Chem. Rev.* **96**, 1223 (1996).

[21] R. W. Zwanzig, *J. Chem. Phys.* **32**, 1173 (1960).

[22] S. Califano, V. Schettino, and N. Neto, in *Lattice Dynamics of Molecular Crystals*, Lecture Notes in Chemistry, Springer, Berlin (1981).

[23] R. E. Allen, G. P. Alldredge, and F. W. de Wette, *Phys. Rev.* **84**, 1648 (1971); 1661 (1971); 1682 (1971).

[24] E. A. Owen and E. L. Yates, *Phil. Mag.* **15**, 487 (1933).

[25] J. P. Toennies, *Europhysics News* **23**, 63 (1992); A. Lahee and J. P. Toennies, *Physics World* (April 1993), p. 61.

[26] G. Benedek, in *Dynamics of Gas-Surface Interaction*, eds. G. Benedek and U. Valbusa, Springer-Verlag, Berlin (1982), p. 227.

[27] M. Wuttig, R. Franchy, and H. Ibach, *Z. Phys. B.* **65**, 71 (1986); *Solid State Comm.* **57**, 445 (1986).

[28] R. D. Levine, *Quantum Mechanics of Molecular Rate Processes*, Oxford University Press (1969).

[29] B. J. Garrison and S. A. Adelman, *J. Chem. Phys.* **67**, 2379 (1977); A. Redondo, Y. Zeiri, J. J. Low, and W. A. Goddard III, *J. Chem. Phys.* **79**, 6410 (1983).

[30] P. Saalfrank and W. H. Miller, *Surf. Sci.* **303**, 206 (1994).

[31] A. F. Voter and J. D. Doll, *J. Chem. Phys.* **82**, 80 (1985).

[32] E. Ganz, S. K. Theiss, L-S. Hwang, and J. Golovchenko, *Phys. Rev. Lett.* **68**, 1567 (1992).

[33] F. Hofmann and J. P. Toennies, *Chem. Rev.* **96**, 1307 (1996).

[34] E. P. Wigner, *Z. Phys. Chem.* **B19**, 203 (1932).

[35] C. T. Chudley and R. J. Elliott, *Proc. Phys. Soc.* London **77**, 353 (1961).

[36] L. van Hove, *Phys. Rev.* **95**, 249 (1954).

[37] F. Hofmann, J. P. Toennies, and J. R. Manson, *Surf. Sci. Lett.* **349**, L184 (1996).

[38] A. Bilic and B. Gumhalter, *Phys. Rev. B* **52**, 12307 (1995); B. Gumhalter and A. Bilic, *Surf. Sci.* **370**, 47 (1997).

[39] A. Brako and D. M. Newns, *Phys. Rev. Lett.* **48**, 1859 (1982).

[40] G. D. Billing, *Appl. Surf. Sci.* **142**, 7 (1999).

[41] D. A. Cruz and M. Xiao, *Chem. Phys. Lett.* **297**, 225 (1998).

[42] P. W. Fowler and J. M. Hutson, *Phys. Rev. B* **33**, 3724 (1986).

[43] J. M. Hutson and P. W. Fowler, *Surf. Sci.* **173**, 337 (1986).

[44] J. G. Lauerdale and D. G. Truhlar, *J. Chem. Phys.* **84**, 1843 (1985); T. N. Truong and D. G. Truhlar, *J. Chem. Phys.* **88**, 6611 (1988); *J. Phys. Chem.* **91**, 6229 (1987).

[45] R. Jaquet and W. H. Miller, *J. Phys. Chem.* **89**, 2139 (1985).

[46] K. Haug, G. Wahnström and H. Metiu, *J. Chem. Phys.* **90**, 540 (1989); K. Haug and H. Metiu, *J. Chem. Phys.* **94**, 3251 (1991).

[47] W. H. Miller, S. D. Schwartz, and J. W. Tromp, *J. Chem. Phys.* **79**, 4889 (1983).

[48] A. P. Graham, A. Menzel, and J. P. Toennies, *J. Chem. Phys.* (in press)

[49] C. W. Skorupka and J. R. Manson, *Phys. Rev. B.* **41**, 9783 (1990).

[50] A. Bilic and B. Gumhalter, *Phys. Rev. B.* **52**, 12307 (1995).

[51] L. M. Hubbard and W. H. Miller, *J. Chem. Phys.* **78**, 1801 (1983); **80**, 5827 (1984).

[52] G. Wolken, Jr., *J. Chem. Phys.* **58**, 3047 (1973).

[53] D. A. Micha, *J. Chem. Phys.* **74**, 2054 (1981).

[54] M. F. Bertino, F. Hofmann, W. Steinhögl, and J. P. Toennies, *J. Chem. Phys.* **105**, 1 (1996).

[55] R. Kubo, M. Yokota, and S. Nakajima, *J. Phys. Soc. Jpn.* **12**, 1203 (1957).

[56] Lichang Wang, Qingfeng Ge, and G. D. Billing, *Surf. Sci.* **301**, 353 (1994).

[57] G. D. Billing, *Comp. Phys. Rep.* **1**, 237 (1984).

[58] G. D. Billing, *J. Chem. Phys.* **99**, 5849 (1993); *Chem. Phys.* **189**, 523 (1994).

[59] J. Böheim and W. Brenig, *Z. Phys. B:- Condensed Matter* **41**, 243 (1981).

[60] D. Secrest and B. R. Johnson, *J. Chem. Phys.* **45**, 4556 (1966).

[61] G. D. Billing, *Chem. Phys.* **51**, 417 (1980); *Comp. Phys. Comm.* **1**, 237 (1984).

[62] G. D. Billing, *Chem. Phys.* **74**, 143 (1983).

[63] K. C. Janda, J. E. Hurst, C. A. Becker, J. P. Cowin, D. J. Auerbach, and L. Wharton, *J. Chem. Phys.* (1980). **72**, 2403 (1980).

[64] G. D. Billing, *Chem. Phys.* **115**, 229 (1987).

[65] R. J. Glauber, *Phys. Rev.* **98**, 1692 (1955).

[66] D. M. Newns, *Surf. Sci.* **154**, 658 (1985).

[67] H.-D. Meyer, *Surf. Sci.* **104**, 117 (1981).

[68] R. Schinke and R. B. Gerber, *J. Chem. Phys.* **82**, 1567 (1985).

[69] B. Gumhalter and A. Bilic, *Surf. Sci.* **370**, 47 (1997).

[70] G. H. Vineyard, *Phys. Rev.* **110**, 999 (1958); P. Schofield, *Phys. Rev. Lett.* **4**, 239 (1960).

[71] L. M. Raff, J. Lorenzen, and B. C. McKoy, *J. Chem. Phys.* **46**, 4265 (1967); J. Lorenzen and L. M. Raff, *J. Chem. Phys.* **49**, 1165 (1968); **52**, 1133 (1970); **74**, 3929 (1981).

[72] L. D. Landau, *Z. Phys. Sov.* **2**, 46 (1932); C. Zener, *Proc. R. Soc.* London, **A137**, 696 (1932); C. Zhu and H. Nakamura, *J. Chem. Phys.* **101**, 10630 (1994); **102**, 7448 (1995).

[73] J. C. Tully and R. K. Preston, *J. Chem. Phys.* **55**, 562 (1971); J. R. Stine and J. T. Muckerman, *J. Chem. Phys.* **65**, 3975 (1976); J. C. Tully, *J. Chem. Phys.* **93**, 1061 (1990).

[74] G. D. Billing, *Int. Revs. Phys. Chem.* **13**, 309 (1994).

[75] G. D. Billing, *Surf. Sci.* **203**, 257 (1988).

[76] B. Gumhalter and A. Bilic, *Surf. Sci.* **370**, 47 (1997).

[77] B. Jackson, *J. Chem. Phys.* **89**, 2473 (1988).

[78] D. R. Reichman and R. J. Silbey, *J. Chem. Phys.* **104**, 1506 (1996).

[79] P. Pechukas, *Phys. Rev. Lett.* **73**, 1060 (1994); R. Alicki, Ibid, **75**, 3020 (1995); P. Pechukas, Ibid, **75**, 3021 (1995).

[80] F. Hofmann, J. P. Toennies, and J. R. Manson, *Surf. Sci. Lett.* **349**, L184 (1996).

[81] J. L. Beeby, *J. Phys. C.* **5**, 3438 (1972).

[82] T. P. Tsien and B. C. Eu, *J. Chem. Phys.* **89**, 3389 (1988).

[83] S. A. Adelman and J. D. Doll, *J. Chem. Phys.* **61** 4242 (1974); **64**, 2375 (1976); S. A. Adelman and J. D. Doll, *Acc. Chem. Res.* **10**, 378 (1977); S. A. Adelman, *Adv. Chem. Phys.* **44**, 143 (1980).

[84] J. C. Tully, *J. Chem. Phys.* **73**, 1975, 6333 (1980); *Ann. Rev. Phys. Chem.* **31**, 319 (1980); M. Shugard, J. C. Tully, and A. Nitzan, *J. Chem. Phys.* **66**, 2534 (1977).

[85] A. E. DePristo, *Surf. Sci.* **141**, 40 (1984).

[86] S. R. De Groot and P. Mazur, *"Non-equilibrium Thermodynamics"*, North-Holland, Amsterdam, (1962).

[87] A. M. Richard and A. E. DePristo, *Surf. Sci.* **134**, 338 (1983).

[88] Y. Georgievskii, M. A. Kozhushner, and E. Pollak, *J. Chem. Phys.* **102**, 6908 (1995).

[89] C-Y. Lee, R. F. Grote, and A. E. DePristo, *Surf. Sci.* **145**, 466 (1984).

[90] R. Zwanzig, *J. Stat. Phys.* **9**, 215 (1973).

[91] Li Hua Yu and Chang-Pu Sun, *Phys. Rev. A* **49**, 592 (1994); Li Hua Yu, *Phys. Rev. A* **54**, 3779 (1996).

[92] Lord Rayleigh, *Proc. London Math.* **17**, 4 (1887).

[93] M. Wuttig, R. Franchy, and H. Ibach, *Z. Phys. B: Condensed Matter* **65**, 71 (1986); *Solid State Comm.* **57**, 445 (1986).

[94] R. E. Langer, *Phys. Rev.* **51**, 669 (1937).

[95] See, e.g., W. P. Reinhardt, *Prog. Theoretical Phys. Suppl.* **116**, 179 (1994); W. P. Reinhardt and I. Dana, *Proc. R. Soc.* London. A **413**, 157 (1987).

[96] P. Ehrenfest, *Phil. Mag.* **33**, 500 (1917).

[97] C. M. Laursen, E. Christoffersen, and G. D. Billing, *Chem. Phys. Lett.* **290**, 275 (1998).

[98] D. A. McQuarrie, *"Statistical Mechanics"*, Harper & Row, New York, (1976).

[99] K. Möhring and U. Smilansky, *Nucl. Phys.* **A338**, 227 (1980); *J. Chem. Phys.* **74**, 4509 (1981).

[100] K. L. Sebastian, *Chem. Phys. Lett.* **81**, 14 (1981).

[101] G. D. Billing, *Chem. Phys.* **70**, 223 (1982).

[102] R. B. Gerber, V. Buch, and M. A. Ratner, *J. Chem. Phys.* **77**, 3022 (1982); R. B. Gerber and M. A. Ratner, *J. Phys. Chem.* **92**, 3252 (1988); R. B. Gerber, R. Kosloff, and M. Berman, *Comp. Phys. Rep.* **5**, 59 (1986).

[103] P. Pechukas and J. C. Light, *J. Chem. Phys.* **44**, 3897 (1966).

[104] S. C. Park and J. M. Bowman, *Chem. Phys. Lett.* **119**, 275 (1985).

[105] R. Brako and D. M. Newns, *J. Phys. C: Solid State Phys.* **14**, 3065 (1981).

[106] B. Jackson, *J. Chem. Phys.* **90**, 140 (1989).

[107] G. D. Billing, *J. Chem. Phys.* **99**, 5849 (1993).

[108] B. Jackson, *J. Chem. Phys.* **92**, 1458 (1990).

[109] G. D. Billing, *Comp. Phys. Rep.* **1**, 237 (1984).

[110] M. W. Cole and G. Vidali, in *Dynamics of Gas-Surface Interaction*, eds. G. Benedek and U. Valbusa, Springer-Verlag, Berlin (1982), pp. 111.

[111] F. London, *Z. Physik* **63**, 245 (1930).

[112] K. T. Tang and J. P. Tonnies, *J. Chem. Phys.* **80**, 3726 (1984).

[113] G. Brusdeylins, R. B. Doak, and J. P. Toennies, *J. Chem. Phys.* **75**, 1784 (1981); W. Y. Leung, J. Z. Larese, and D. R. Frankl, *Surf. Sci.* **145**, L398 (1984).

[114] H. Chow, *Surf. Sci.* **66**, 2221 (1977).

[115] P. Debye, *Ann. Physik.* **39**, 789 (1912).

[116] P. W. Bridgman, *Proc. Am. Acad. Sci.* **58**, 182 (1922).

[117] G. Brusdeylins, R. B. Doak, and J. P. Toennies, *Phys. Rev. Lett.* **44**, 1417 (1980).

[118] F. Hofmann and J. P. Toennies, *Chem. Rev.* **96**, 1307 (1996).

[119] K. L-Poljanec, M. Hodoscek, D. Lovric, and B. Gumhalter, *Surf. Sci.* **251**, 706 (1991).

[120] G. Ihm, M. W. Cole, F. Toigo, and G. Scoles, *J. Chem. Phys.* **87**, 3995 (1987).

[121] J. A. Barker and C. T. Rettner, *J. Chem. Phys.* **97**, 5844 (1992).

[122] *CRC Handbook of Chemistry and Physics*, CRC Press, London, (1994–95).

[123] J. C. Polanyi, *Accts. Chem. Res.* **5**, 161 (1972).

[124] G. D. Billing, *Comp. Phys. Rep.* **12**, 383 (1990).

[125] R. B. Gerber, L. H. Beard, and D. J. Kouri, *J. Chem. Phys.* **74**, 4709 (1981).

[126] B. R. Johnson, *J. Comp. Phys.* **13**, 445 (1973).

[127] R. B. Gerber, A. T. Yinnon, and J. N. Murrell, *Chem. Phys.* **31**, 1 (1978); R. Schinke, *Chem. Phys. Lett.* **87**, 438 (1982); Y. Shima and M. Baer, *J. Chem. Phys.* **83**, 5250 (1985); M. Baer, *J. Chem. Phys.* **81**, 4526 (1984); *Chem. Phys. Lett.* **116**, 439 (1985).

[128] T. R. Proctor, D. J. Kouri, and R. B. Gerber, *J. Chem. Phys.* **80**, 3845 (1984).

[129] D. J. Kouri, in *Atom-Molecule Collision Theory: A Guide for the Experimentalist*, ed. R. B. Bernstein, Plenum Press, New York, (1979), p, 301.

[130] G. D. Billing and K. V. Mikkelsen, *Introduction to Molecular Dynamics and Chemical Kinetics*, J Wiley, New York (1996).

[131] H. Eyring and M. Polanyi, *Z. Physik. Chem.* **B12**, 279 (1931); F. London, *Z. Elektrochem.* **35**, 552 (1979); S. Sato, *J. Chem. Phys.* **23**, 592, 2465 (1955).

[132] G. Herzberg, *Spectra of Diatomic Molecules*, Van Nostrand, Princeton, NJ (1950); K. P. Huber and G. Herzberg, *Molecular Spectra and Molecular Structure, IV. Constants of Diatomic Molecules*, Van Nostrand, New York (1979).

[133] J.-H. Lin and B. J. Garrison, *J. Chem. Phys.* **80**, 2904 (1984).

[134] M. Person, J. Strömquist, L. Bengtsson, B. Jackson, D. V. Shalashilin, and B. Hammer, *J. Chem. Phys.* **110**, 2240 (1999).

[135] J. H. McCreery and G. Wolken, Jr., *J. Chem. Phys.* **63**, 2340 (1975).

[136] G. D. Billing, *Chem. Phys.* **86**, 349 (1984).

[137] A. M. Lahee, J. P. Toennies, and Ch. Wöll, *Surf. Sci.* **177**, 371 (1986).

[138] A. P. Graham, *J. Chem. Phys.* **109**, 9583 (1998).

[139] H. Froitzheim and M. Schulze, *Surf. Sci.* **211/212**, 837 (1989).

[140] D. F. Ogletree, M. A. van Hove, and G. A. Somarjai, *Surf. Sci.* **173**, 351 (1986).

[141] F. O. Ellison, *J. Am. Chem. Soc.* **85**, 3540 (1963); P. A. Kuntz, in *Atom-Molecule Collision Theory*, ed. R. B. Bernstein, Plenum, New York (1979).

[142] E. Clementi and C. Roetti, *At. Data Nucl. Data Tables* **4**, 177 (1974).

[143] A. D. McLean and R. S. McLean, *At. Data Nucl. Data Tables* **26**, 197 (1981).

[144] N. Govind, Y. A. Wang, A. J. R. da Silva, and E. A. Carter, *Chem. Phys. Lett.* **295**, 129 (1998).

[145] E. Shustorovich and H. Sellers, *Surf. Sci. Rep.* **31**, 1 (1998).

[146] M. S. Daw and M. I. Baskes, *Phys. Rev. Lett.* **50**, 1285 (1983); *Phys. Rev. B* **29**, 6443 (1984).

[147] S. M. Foiles, *Phys. Rev. B.* **48**, 4287 (1993).

[148] M. I. Baskes, *Phys. Rev. B* **46**, 2727 (1992).

[149] G. Haase, M. Asscher, and R. Kosloff, *J. Chem. Phys.* **90**, 3346 (1989); L. Romm, G. Katz, R. Kosloff, and M. Asscher, *J. Phys. Chem.* **101**, 2213 (1997).

[150] A. Kara and A. E. DePristo, *Surf. Sci.* **193**, 437 (1988).

[151] A. Kara and A. E. DePristo, *J. Chem. Phys.* **88**, 2033, 5240 (1988).

[152] P. Siegbahn, M. R. A. Blomberg, I. Panas, and U. Wahlgren, *Theor. Chem. Acta* **75**, 143 (1988).

[153] H. Ibach, *Surf. Sci.* **66**, 56 (1977).

[154] J. W. Gadzuk and S. Holloway, *J. Chem. Phys.* **84**, 3502 (1986); J. W. Gadzuk, *Ann. Rev. Phys.* **39**, 395 (1988).

[155] G. D. Billing, A. Guldberg, N. E. Henriksen, and F. Y. Hansen, *Chem. Phys.* **147**, 1 (1990); N. E. Henriksen, F. Y. Hansen, and G. D. Billing, *Chem. Phys. Lett.* **244**, 350 (1995).

[156] G. D. Billing and K. V. Mikkelsen, *Advanced Molecular Dynamics and Chemical Kinetics*, Wiley, New York (1997).

[157] P. J. Van Den Hoek and E. J. Baerends, *Surf. Sci. Lett.* **221**, 791 (1989); Lichang Wang and G. D. Billing, *Chem. Phys.* **224**, 65 (1997).

[158] B. H. Choi and R. T. Poe, *J. Chem. Phys.* **83**, 1330 (1985).

[159] L. Pesce and P. Saalfrank, *Chem. Phys.* **219**, 43 (1997).

[160] R. Kosloff, M. A. Ratner, and B. Davis, *J. Chem. Phys.* **106**, 7036 (1997).

[161] D. A. Micha and Z. Yi, *J. Chem. Soc. Faraday Trans.* **93**, 969 (1997).

[162] G. Lindblad, *Comm. Math. Phys.* **48**, 119 (1976).

[163] A. Kossakowski, *Rep. Math. Phys.* **3**, 247 (1972).

[164] C. Cerjan and R. Kosloff, *Phys. Rev. B* **34**, 3832 (1986); P. Saalfrank and R. Kosloff, *J. Chem. Phys.* **105**, 2441 (1996).

[165] B. Davies, Commun. *Math. Phys.* **39**, 91 (1976).

[166] B. Jackson, *Chem. Phys. Lett.* **270**, 484 (1997).

[167] A. G. Redfield, *Adv. Magn. Reson.* **1**, 1 (1966).

[168] R. B. Gerber, R. Kosloff, and M. Berman, *Comp. Phys. Rep.* **5**, 59 (1986).

[169] M. D. Feit, J. R. Fleck, and A. Steiger, *J. Comp. Phys.* **47**, 412 (1982).

[170] G. Katz and R. Kosloff, *J. Chem. Phys.* **103**, 9475 (1995).

[171] P. Pechukas and P. Hänggi, *Phys. Rev. Lett.* **73**, 2772 (1994); C. R. Doering and J. C. Gadoua, *Phys. Rev. Lett.* **69**, 2318 (1992); L. H. Yu, *Phys. Rev. A.* **54**, 3779 (1996); L. H. Yu and C-P. Sun, *Phys. Rev. A.* **49**, 592 (1994).

[172] H. A. Kramers, *Physica* (Utrecht) **7**, 284 (1940).

[173] N. Makri and W. H. Miller, *J. Chem. Phys.* **87**, 5781 (1987); G. D. Billing and G. Jolicard, *Chem. Phys. Lett.* **221**, 75 (1994).

[174] See, e.g., H. Eyring, J. Walter, and G. E. Kimball, *Quantum Chemistry*, Wiley, New York (1965).

[175] P. Haochang, T. C. M. Horn, and A. W. Kleyn, *Phys. Rev. Lett.* **57**, 3035 (1986); P. H. F. Reijnen and A. W. Kleyn, *Chem. Phys.* **139**, 489 (1989); M. E. M. Spruit and A. W. Kleyn, *Chem. Phys. Lett.* **159**, 342 (1989); Lichang Wang and G. D. Billing, *Chem. Phys.* **224**, 65 (1997).

[176] R. Kosloff and O. Citri, *Faraday Discuss.* **96**, 175 (1993).

[177] S. M. Foiles, M. L. Baskes, and M. S. Daw, *Phys. Rev. B* **33**, 7983 (1986).

[178] J. A. White and D. M. Bird, *Chem. Phys. Lett.* **213**, 422 (1993).

[179] G. D. Billing, *J. Phys. Chem.* **99**, 15378 (1995).

[180] J. A. White, D. M. Bird, M. Payne, and I. Stich, *Phys. Rev. Lett.* **73**, 1404 (1994).

[181] B. Hammer, M. Scheffler, K. W. Jacobsen, and J. K. Nørskov, *Phys. Rev. Lett.* **73**, 1400 (1994).

[182] N. D. Lang and A. R. Williams, *Phys. Rev. Lett.* **34**, 531 (1975); **37**, 212 (1976); *Phys. Rev. B* **18**, 616 (1978).

[183] B. Hammer, K. W. Jacobsen, and J. K. Nørskov, *Phys. Rev. Lett.* **70**, 3971 (1993).

[184] G. Wiesnekker, G. J. Kroes, E. J. Baerends, and R. C. Mowrey, *J. Chem. Phys.* **102**, 3873 (1995).

[185] B. Hammer and M. Scheffler, *Phys. Rev. Lett.* **74**, 3487 (1995).

[186] D. M. Bird, L. J. Clarke, M. C. Payne, and I. Stich, *Chem. Phys. Lett.* **212**, 518 (1993).

[187] J. A. White, D. M. Bird, and M. C. Payne, *Phys. Rev. B* **53**, 1667 (1996).

[188] P. Kratzer, B. Hammer, and J. K. Nørskov, *Phys. Rev. B* **51**, 13432 (1995).

[189] See, e.g., P. Madhaven and J. L. Whitten, *J. Chem. Phys.* **77**, 2673 (1982); A. Kokalj, A. Lesar, and M. Hodoscek, *Chem. Phys. Lett.* **268**, 43 (1997).

[190] B. Hammer and J. K. Nørskov, *Phys. Rev. Lett.* **79**, 4441 (1997).

[191] J. J. Mortensen, B. Hammer, and J. K. Nørskov, *Surf. Sci.* **414**, 315 (1998).

[192] E. V. Stefanovich and T. N. Truong, *Chem. Phys. Lett.* **299**, 622 (1999).

[193] M. A. van Daelen, M. Neurock, and R. A. van Santen, *Surf. Sci.* **417**, 247 (1998).

[194] J. N. Murrel, S. Carter, S. C. Farantos, P. Huxley and A. J. C. Varandas, *Molecular Potential Energy Functions*, Wiley, New York (1984).

[195] G. Makov and M. C. Payne, *Phys. Rev. B* **51**, 4014 (1995).

[196] G. D. Billing and M. Cacciatore, *Chem. Phys.* **103**, 137 (1986).

[197] L. Verlet, *Phys. Rev.* **159**, 98 (1967).

[198] M. P. Teter, M. C. Payne, and D. C. Allan, *Phys. Rev. B* **40**, 12255 (1989); J. M. Milliam and G. E. Scuseria, *J. Chem. Phys.* **106**, 5569 (1997).

[199] M. C. Payne, M. P. Teter, D. C. Allan, T. A. Arias, and J. D. Joannopoulos, *Rev. Mod. Phys.* **64**, 1045 (1992).

[200] J. E. Müller, *Surf. Sci.* **272**, 45 (1992).

[201] N. H. March and A. M. Murray, *Proc. Roy. Soc. London Ser. A.* **261**, 119 (1961); *Phys. Rev.* **120**, 830 (1960).

[202] J. Lindhard, *Kgl. Danske Mat. Medd.* **28**, 8 (1954).

[203] J. M. C. Scott, *Phil. Mag.* **43**, 859 (1952).

[204] D. J. Lacks and R. G. Gordon, *J. Chem. Phys.* **100**, 4446 (1994).

[205] A. I. M. Rae, Chem. *Phys. Lett.* **18**, 574 (1973); *Mol. Phys.* **29**, 467 (1975).

[206] N. H. March, in *Theory of the Inhomogeneous Electron Gas*, eds. S. Lundqvist and N. H. March, Plenum Press, New York (1983), pp. 1–78.

[207] R. G. Gordon and Y. S. Kim, *J. Chem. Phys.* **56**, 3122 (1972).

[208] C. Nyeland and J. P. Toennies, *Chem. Phys.* **188**, 205 (1994).

[209] T. N. Truong, D. G. Truhlar, and B. C. Garrett, *J. Phys. Chem.* **93**, 8227 (1989).

[210] N. Esbjerg and J. K. Nørskov, *Phys. Rev. Lett.* **45**, 807 (1980); J. K. Nørskov and N. D. Lang, *Phys. Rev. B* **21**, 2131 (1980); J. D. Kress and A. E. DePristo, *J. Chem. Phys.* **87**, 4700 (1987); K. W. Jacobsen, J. K. Nørskov, and M. J. Puska, *Phys. Rev. B* **35**, 7423 (1987); T. J. Raeker and A. E. DePristo, *Surf. Sci.* **248**, 134 (1991).

[211] P. Hohenberg and W. Kohn, *Phys. Rev.* **136**, B864 (1964).

[212] R. G. Parr and W. Yang, *Density-Functional Theory of Atoms and Molecules*, Oxford University Press, Oxford (1989); R. O. Jones and O. Gunnarsson, *Rev. Mod. Phys.* **61**, 689 (1989); M. C. Payne, M. P. Teter, D. C. Allan, T. A. Arias, and J. D. Joannopolous, *Rev. Mod. Phys.* **64**, 1045 (1992); *Theoretical and Computational Chemistry: Recent Developments and Application of Modern Density Functional Theory*, Ed. J. M. Seminario, Elsevier Amsterdam, (1996).

[213] P. P. Ewald, *Ann. Phys.* **54**, 519 (1917).

[214] M. Levy, *Proc. Natl. Acad. Sci.* **76**, 6062 (1979).

[215] W. Kohn and L. J. Sham, *Phys. Rev.* **136**, 1133 (1965).

[216] E. P. Wigner, *Phys. Rev.* **46**, 1002 (1934); *Trans. Faraday. Soc.* **34**, 678 (1938).

[217] E. Fermi, *Z. Phys.* **48**, 73 (1928).

[218] L. H. Thomas, *Proc. Camb. Phil. Soc.* **23**, 542 (1926).

[219] V. L. Morruzzi, J. F. Janak, and A. R. Williams, *Calculated Electronic Properties of Metals*, Pergamon, New York (1978).

[220] M. Gell-Mann and K. A. Brueckner, *Phys. Rev.* **106**, 364 (1957).

[221] P. Nozieres and D. Pines, *Phys. Rev.* **111**, 442 (1958).

[222] O. Gunnarson and B. I. Lundquist, *Phys. Rev. B* **13**, 4274 (1976).

[223] V. von Barth and L. Hedin, *J. Phys. C.* **5**, 1629 (1972).

[224] S. J. Vosko, L. Wilk, and M. Nusair, *Can. J. Phys.* **58**, 1200 (1980).

[225] J. P. Perdew and Y. Wang, *Phys. Rev. B* **45**, 13244 (1992).

[226] E. K. U. Gross, M. Petersilka, and T. Grabo, in *Chemical Applications of Density-Functional Theory*, eds. B. B. Laird, R. B. Ross, and T. Ziegler, ACS Symposium Series 629, Washington DC (1996), p. 42.

[227] J. P. Perdew and Y. Wang, *Phys. Rev. A* **38**, 8800 (1986).

[228] J. P. Perdew, K. Burke, and M. Ernzerhof, *Phys. Rev. Lett.* **77**, 3865 (1996).

[229] C. Lee, W. Yang, and R. G. Parr, *Phys. Rev. B* **37**, 785 (1988).

[230] A. D. Becke, *Phys. Rev. A* **38**, 3098 (1988).

[231] B. Y. Tong and L. J. Sham, *Phys. Rev.* **144**, 1 (1966).

[232] Y-M. Juan, E. Kaxiras, and R. G. Gordon, *Phys. Rev. B* **51**, 9521 (1995).

[233] T. van Voorhis and G. E. Scuseria, *Mol. Phys.* **92**, 601 (1997); *J. Chem. Phys.* **109**, 400 (1998).

[234] R. Gaspar, *Acta Phys. Hung.* **3**, 263 (1954).

[235] P. J. H. Denteneer and W. van Haeringen, *J. Phys. C.* **18**, 4127 (1985); G. B. Bachelet, D. R. Hamann, and M. Schlüter, *Phys. Rev. B.* **26**, 4199 (1982); G. Oriz and P. Ballone, *Phys. Rev. B* **43**, 6376 (1991).

[236] B. G. Johnson, P. M. W. Gill, and J. A. Pople, *J. Chem. Phys.* **98**, 5612 (1993).

[237] W. H. Press, B. P. Flannery, S. A. Teukolsky, and W. T. Vetterling, *Numerical Recipes*, Cambridge University Press, Cambridge (1986).

[238] E. van Lenthe, E. J. Baerends, and J. G. Snijders, *J. Chem. Phys.* **101**, 9783 (1994).

[239] R. Car and M. Parinello, *Phys. Rev. Lett.* **55**, 2471 (1985).

[240] A. P. Smith, J. K. Wiggs, H. Jonsson, H. Yan, L. R. Corrales, P. Nachtigall, and K. D. Jordan, *J. Chem. Phys.* **102**, 1044 (1995).

[241] J. Wiggs and H. Jonsson, *Comp. Phys. Comm.* **87**, 319 (1995).

[242] E. Deumens, A. Diz, R. Longo, and Y. Örhn, *Rev. Mod. Phys.* **66**, 917 (1994); M. Ben-Nun and R. D. Levine, *J. Phys. Chem.* **100**, 7884 (1996).

[243] F. London, *Z. Electrochem.* **35**, 552 (1929).

[244] J. K. Nørskov, *Rep. Prog. Phys.* **50**, 1253 (1990).

[245] See, e.g., J. C. Slater and N. H. Frank, *Electromagnetism*, McGraw-Hill, New York, (1947).

[246] A. D. Becke and R. M. Dickson, *J. Chem. Phys.* **89**, 2993 (1988).

[247] D. M. Newns, *Phys. Rev.* **178**, 1123 (1968).

[248] S. Raimes, *The Wave Mechanics of Electrons in Metals*, North-Holland, Amsterdam (1963); D. Pines, *Elementary Excitations in Solids*, Benjamin/Cummings, Reading (1963).

[249] M. Persson and B. Hellsing, *Phys. Rev. Lett.* **49**, 662 (1982); B. Hellsing and M. Persson, *Physica Scripta* **29**, 360 (1984).

[250] P. W. Anderson, *Phys. Rev. Lett.* **18**, 1049 (1967).

[251] J. W. Gadzuk, *Phys. Rev. B.* **24**, 1866 (1981).

[252] G. P. Brivio and T. B. Grimley, *J. Phys. C.* **10**, 2351 (1977); *Surf. Sci.* **89**, 226 (1979).

[253] D. Kumamato and R. Silbey, *J. Chem. Phys.* **75**, 5164 (1981).

[254] Z. Kirson, R. B. Gerber, and A. Nitzan, *Surf. Sci.* **124**, 279 (1983).

[255] O. Gunnarson and K. Schönhammer, *Phys. Rev. B. Condens. Matter* **25**, 2514 (1982).

[256] S. Efrima, C. Jedrzejek, K. F. Freed, E. Hood, and H. Metiu, *J. Chem. Phys.* **79**, 2436 (1983); G. E. Korzeniewski, E. Hood, and H. Metiu, *J. Chem. Phys.* **80**, 6274 (1984); K. M. Leung, G. Schön, P. Rudolph, and H. Metiu, *J. Chem. Phys.* **81**, 3307 (1984).

[257] G. D. Billing, *Chem. Phys. Lett.* **290**, 150 (1998).

[258] Y. Li and G. Wahnström, *Phys. Rev. B* **51**, 12233 (1995).

[259] B. N. J. Persson and J. W. Gadzuk, *Surf. Sci. Lett.* **410**, (1998), 779.

[260] R. Baer, Y. Zeiri, and R. Kosloff, *Phys. Rev. B* **55**, 10952 (1997).

[261] J. Bardeen, *Phys. Rev.* **49**, 653 (1936).

[262] G. D. Billing, *Chem. Phys.* **116**, 269 (1987).

[263] E. Blaisten-Barojas and J. W. Gadzuk, *J. Chem. Phys.* **97**, 862 (1992).

[264] M. Head-Gordon and J. C. Tully, *J. Chem. Phys.* **96**, 3939 (1992).

[265] M. Morin, N. J. Levinos, and A. L. Harris, *J. Chem. Phys.* **96**, 3959 (1992).

[266] F. Bloch, *Zeit. für Physik* **52**, 555 (1928).

[267] S. K. Mengel and G. D. Billing, *J. Phys. Chem.* **101**, 10781 (1997).

[268] G. D. Billing, *J. Chem. Phys.* **110**, 5526 (1999); *Chem. Phys. 242*, 341 (1999).

[269] L. Jeloaica and V. Sidis, *Chem. Phys. Lett.* **300**, 157 (1999).

[270] D. A. McCormack, G-J. Kroes, E-J. Baerends, and R. C. Mowrey, Faraday Discuss. *110*, 267 (1998); D. A. McCormack and G-J. Kroes, Phys. Chem. Chem. Phys. *1*, 1359 (1999).

[271] D. H. Zhang and J. Z. H. Zhang, in *Dynamics of Molecules and Chemical Reactions*, eds. R. E. Wyatt and J. Z. H. Zhang, Marcel Dekker, New York (1996). *Faraday Discuss.* **110**, 267 (1998).

[272] M. Kay, G. R. Darling, and S. Holloway, *J. Chem. Phys.* **108**, 4614 (1998).

[273] Y. Shima and M. Baer, *J. Chem. Phys.* **83**, 5250 (1985).

[274] S. S. Fu and G. A. Somarjai, *Surf. Sci.* **262**, 68 (1992).

[275] J. Yoshihara and C. T. Campbell, *J. Catal.* **161**, 776 (1996).

[276] V. P. Zhdanov and B. Kasemo, *Surf. Sci. Rep.* **29**, 31 (1997).

[277] See, e.g., O. Kortlüke and W. von Niessen, *Surf. Sci.* **401**, 185 (1998); V. P. Zhdanov and B. Kasemo, *Surf. Sci.* **405**, 27 (1998).

[278] G. Ertl and H-J. Freund, Catalysis and surface science, in *Physics Today* (Jan. 1999), p. 32.

[279] F. Besenbacher, I. Chorkendorff, B. S. Clausen, B. Hammer, A. M. Molenbroek, J. K. Nørskov, and I. Stensgaard, *Science* **279**, 1913 (1998).

[280] P. Pechukas, in *Dynamics of Molecular Collisions*, ed. W. H. Miller, Plenum Press, New York (1976).

[281] R. I. Masel, *Principles of Adsorption and Reaction on Solid Surfaces*, Wiley, New York (1996), p. 305.

[282] B. F. Hansen and G. D. Billing, *Surf. Sci. Lett.* **373**, 333 (1997).

[283] B. C. Garrett and D. G. Truhlar, *J. Phys. Chem.* **83**, 1052, 1079 (1979); *J. Chem. Phys.* **72**, 3460 (1980); B. C. Garrett, D. G. Truhlar, and R. S. Grev, in *Potential Energy Surfaces and Dynamics Calculations*, ed. D. G. Truhlar, Plenum Press, New York (1981), p. 587.

[284] E. T. Whittaker and G. N. Watson, *A Course of Modern Analysis*, Cambridge University Press (1973), p. 265.

[285] I. S. Gradshteyn and I. M. Ryzhik, *Tables of Integrals, Series, and Products*, Academic Press, New York (1965).

ANSWERS TO EXERCISES

Chapter 1

1.1: The rate of adsorption is $v_{ads} = k_{ads}[A](1 - \theta)$ where θ is the fraction of the surface which is covered. The desorption rate is $v_{des} = k_{des}\theta$ and hence $\theta = K[A]/(1 + K[A])$ where $K = k_{ads}/k_{des}$.

1.2: We get $K = 4.9$ atm^{-1} or 120 l/mol.

1.3: We introduce the following elementary steps

$$H_2(g) + * \rightleftharpoons H_2^* \tag{AN.1}$$

$$H_2^* + * \rightleftharpoons 2H^* \tag{AN.2}$$

where * denotes a free site and H^* and adsorbed hydrogen atom. Introducing the two equilibrium constants

$$\frac{[H_2^*]}{[S]p_{H_2}} = K_1 \tag{AN.3}$$

$$\frac{[H^*]^2}{[H_2^*][S]} = K_2 \tag{AN.4}$$

where $[S]$ is the concentration of free sites and p_{H_2} the pressure of hydrogen, we have

$$\frac{[H^*]}{[S]} = \sqrt{K_1 K_2 p_{H_2}} \tag{AN.5}$$

Introducing the coverage as $\theta = [H^*]/[S_0]$ and $[S] = [S_0] - [H^*]$ we obtain the desired result.

1.4: The fit to the isotherm derived in exercise 1.3 is not very good. Actually a power of $p_{H_2}^{3/2}$ gives a much better fit (see ref. [281]).

1.5: We get 7.6 10^{-18} at 300 K and 2.4 10^{-13} cm^3/sec. at 1000 K.

1.6: $z^* = 2.9$ Å.

1.7: The average value is obtained as $(4\mu)/(3(\mu+1)^2)$ since

$$\int_0^{\pi/2} d\theta \sin\theta \cos^2\theta / \int_0^{\pi/2} d\theta \sin\theta = 1/3.$$

1.8: By equating eqs. (A.11) and (A.22) we obtain

$$k_{ER}/k_{LH} = \frac{P_A}{RT}\Pi_{i=1}^{3}\frac{1}{2\sinh(\hbar\omega_i^*/2k_BT)}\frac{h^3}{\theta_A(2\pi m_A k_B T)^{3/2}} P_{steric}\exp(E_b/k_BT) \qquad \text{(AN.6)}$$

where $E_b = E_A - E_A^*$ is the binding energy and θ_A the surface coverage coefficient $\theta_A = [A]^*A$. Assuming a pressure of 1 torr and $\theta_A = 1$ we get the ratio 3.4 $10^{-7}\exp(E_b/k_BT)$. Thus $E_b = 49.5$ kJ/mol binding energy or more will tend to favor the ER mechanism, if $P_{steric} = 1$.

1.9: The flux on a surface area is given as

$$j = \frac{N}{V}\int_0^{\infty} dv_z f(v_z)v_z \qquad \text{(AN.7)}$$

where N is the number of molecules in the volume V. Using the one-dimensional velocity distribution (eq. (1.3)) we get $j = p/\sqrt{2\pi m k_B T}$ where p is the gas pressure. Insertion in this expression gives $j = 1.09\ 10^{16}$ molecuels/cm^{2}sec. One cm^2 is covered in 0.15 sec.

Chapter 2

2.1: Introduce a coordinate system such that

$$\mathbf{a} = a(1, 0, 0)$$
$$\mathbf{b} = b(\cos\gamma, \sin\gamma, 0)$$
$$\mathbf{c} = c(\cos\beta, (\cos\alpha - \cos\beta\cos\gamma)/\sin\gamma, c_3) \qquad \text{(AN.8)}$$

where c_3 is obtained as

$$c_3^2 = 1 - \cos^2\beta - (\cos\alpha - \cos\beta\cos\gamma)^2)/\sin^2\gamma \qquad \text{(AN.9)}$$

which follows from the normalization and the definition of the angles. Inserting these expressions in $V = \mathbf{a}\cdot\mathbf{b}\times\mathbf{c}$, we obtain the desired expression for the volume of a unit cell. In order to obtain the distance between the planes, we note that the planes cut the axes at $\mathbf{a}/h$, $\mathbf{b}/k$, and $\mathbf{c}/l$. Thus we need to find the perpendicular distance from the origin (0,0,0) to the plane. Two vectors in the plane are

$$\mathbf{x} = \frac{\mathbf{c}}{l} - \frac{\mathbf{a}}{h} \tag{AN.10}$$

$$\mathbf{y} = \frac{\mathbf{b}}{k} - \frac{\mathbf{a}}{h} \tag{AN.11}$$

and hence $\mathbf{x} \times \mathbf{y}$ is a vector perpendicular to the plane. The distance d_{hkl} can be obtained as

$$d_{hkl} = \frac{a}{h}\cos\Delta \tag{AN.12}$$

where Δ is the angle between $\mathbf{a}$ and the surface normal, i.e.,

$$\cos\Delta = \frac{\mathbf{a} \cdot \mathbf{x} \times \mathbf{y}}{a|\mathbf{x} \times \mathbf{y}|} \tag{AN.13}$$

Inserting the above expressions for $\mathbf{x}$ and $\mathbf{y}$ we obtain

$$\mathbf{a} \cdot \mathbf{x} \times \mathbf{y} = \frac{V}{lk} \tag{AN.14}$$

and hence

$$d_{hkl} = \frac{V}{hkl|\mathbf{x} \times \mathbf{y}|} \tag{AN.15}$$

Inserting the expressions for the vectors $\mathbf{a}$, $\mathbf{b}$, and $\mathbf{c}$ we can show that

$$|\mathbf{c} \times \mathbf{b}|^2 = b^2c^2\sin^2\alpha \tag{AN.16}$$

$$|\mathbf{a} \times \mathbf{b}|^2 = a^2b^2\sin^2\gamma \tag{AN.17}$$

$$|\mathbf{c} \times \mathbf{a}|^2 = a^2c^2\sin^2\beta \tag{AN.18}$$

$$\mathbf{c} \times \mathbf{b} \cdot \mathbf{c} \times \mathbf{a} = abc^2(\cos\gamma - \cos\beta\cos\alpha) \tag{AN.19}$$

$$\mathbf{c} \times \mathbf{b} \cdot \mathbf{a} \times \mathbf{b} = ab^2c(\cos\beta - \cos\gamma\cos\alpha) \tag{AN.20}$$

$$\mathbf{c} \times \mathbf{a} \cdot \mathbf{a} \times \mathbf{b} = -a^2bc(\cos\alpha - \cos\beta\cos\gamma) \tag{AN.21}$$

which finally gives the desired result.

Alternatively use eq. (2.21) with $\mathbf{l} = (h, k, l)$ and hence

$$|\mathbf{l} \cdot \mathbf{B}|^2 = h^2\mathbf{b}_1 \cdot \mathbf{b}_1 + k^2\mathbf{b}_2 \cdot \mathbf{b}_2 + l^2\mathbf{b}_3 \cdot \mathbf{b}_3 + 2hk\mathbf{b}_1 \cdot \mathbf{b}_2 + 2hl\mathbf{b}_1 \cdot \mathbf{b}_3 + 2kl\mathbf{b}_2 \cdot \mathbf{b}_3 \tag{AN.22}$$

Using the fact that

$$|\mathbf{b}_1|^2 = \frac{(2\pi)^2}{V^2}|(\mathbf{a}_2 \times \mathbf{a}_3)|^2 = 4\pi^2 b^2 c^2 \sin^2\alpha / V^2 \tag{AN.23}$$

and

$$\mathbf{b}_1 \cdot \mathbf{b}_2 = \frac{(2\pi)^2}{V^2}(\mathbf{a}_2 \times \mathbf{a}_3 \cdot \mathbf{a}_3 \cdot \mathbf{a}_1 = 4\pi^2 abc^2(\cos\alpha\cos\beta - \cos\gamma)) \tag{AN.24}$$

etc., we can obtain the expression. Here we have used that $\mathbf{a} \times \mathbf{b} \cdot \mathbf{c} \times \mathbf{d} = (\mathbf{a} \cdot \mathbf{c})(\mathbf{b} \cdot \mathbf{d}) - (\mathbf{a} \cdot \mathbf{d})(\mathbf{b} \cdot \mathbf{c})$.

2.2: $\lambda = 1.02$ eV

2.3: Consider Figure 2.4. We see that the path difference is $\Delta = x - y = \mathbf{r} \cdot \mathbf{e}_2 - \mathbf{r} \cdot \mathbf{e}_1 = N\lambda$.

The Laue conditions show that the vector $\mathbf{s} = \mathbf{e}_2 - \mathbf{e}_1$ is perpendicular to the plane (h, k, l) and distance between two lattice planes is

$$d = \frac{\mathbf{a} \cdot \mathbf{s}}{|\mathbf{s}|} = \frac{\lambda h}{s} = \frac{\lambda h}{2 \sin\theta} \tag{AN.25}$$

where we have used the fact that the angle between $\mathbf{e}_1$ and $\mathbf{e}_2$ is $\pi - 2\theta$ and hence $\mathbf{e}_1 \cdot \mathbf{e}_2 = -\cos(\pi - 2\theta)$. Thus $|\mathbf{s}| = \sqrt{2 - 2\mathbf{e}_1 \cdot \mathbf{e}_2} = \sqrt{2(1 - \cos(2\theta))} = 2 \sin\theta$.

2.4: The vector $(1,1,1)/\sqrt{3}$ is perpendicular to the cutting plane and should be rotated to (0,0,1). The rotational angles are defined by: $\cos\alpha = 1/\sqrt{2}$, $\sin\alpha = -1/\sqrt{2}$, $\cos\beta = 1/\sqrt{3}$, and $\sin\beta = -2/\sqrt{6}$. The rotation matrix is then given by eq. (2.7) and the basis vectors by the columns of this matrix.

2.5: Using the fact that $d = a/\sqrt(h^2 + k^2 + l^2)$ we get from eq. (2.3): $\theta = 21.6°$, $\theta = 25.2°$, $\theta = 37.0°$, and $\theta = 44.9°$.

2.6: The matrix is

$$\begin{pmatrix} \frac{1}{\sqrt{2}} & -\frac{1}{\sqrt{2}} & 0 \\ 0 & 0 & 1 \\ \frac{1}{\sqrt{2}} & \frac{1}{\sqrt{2}} & 0 \end{pmatrix} \tag{AN.26}$$

2.7: By summing over lattice points within a 2D sphere of the surface with radius R and dividing by πR^2 we get the number of atoms per unit area. Consider, for instance, the (111) surface of an FCC crystal; we have the coordinates of the atoms given by

$$\begin{pmatrix} x \\ y \\ z \end{pmatrix} = \frac{a}{2} \begin{pmatrix} \frac{1}{\sqrt{2}} & -\frac{1}{\sqrt{2}} & 0 \\ \frac{1}{\sqrt{6}} & \frac{1}{\sqrt{6}} & -\frac{2}{\sqrt{6}} \\ \frac{1}{\sqrt{3}} & \frac{1}{\sqrt{3}} & \frac{1}{\sqrt{3}} \end{pmatrix} \begin{pmatrix} n_2 + n_3 \\ n_1 + n_3 \\ n_1 + n_2 \end{pmatrix}$$

$$= a \begin{pmatrix} \frac{1}{2\sqrt{2}}(n_1 - n_2) \\ \frac{1}{2\sqrt{6}}(n_1 + n_2 - 2n_3) \\ \frac{1}{\sqrt{3}}(n_1 + n_2 + n_3) \end{pmatrix} \quad \text{(AN.27)}$$

Requiring $z = 0$ we have $n_3 = -n_1 - n_2$, and hence $x = (a)(2\sqrt{2})(n_1 - n_2)$ and $y = (3a)(2\sqrt{6})(n_1 + n_2)$. The sum over n_1 and n_2 is converted to an integral over x and y and hence the density is the inverse Jacobian for the transformation or $4/(\sqrt{3}a^2)$. For the (100) cut we get, in a similar manner, $2/a^2$, and for the (110) surface $2\sqrt{2}/a^2$. For a simple cubic lattice, we can, by using the result of exercise 2.6, obtain $1/\sqrt{3}a^2$, $1/a^2$, and $1/a^2$ for the (111), (100), and (110) surfaces, respectively.

2.8: By using eq. (2.10) we obtain $\tan\alpha = -h/k$ and $\cos\beta = l/\sqrt{h^2 + k^2 + l^2}$.

2.9: The equation of motion is

$$m\frac{d^2\xi_{jk}}{dt^2} = F(\xi_{j+1,k} + \xi_{j-1,k} - 2\xi_{jk} + \xi_{j,k+1} + \xi_{j,k-1} - 2\xi_{jk}) \quad \text{(AN.28)}$$

and with the substitution $\xi_{jk} = \exp(i(ak_x j + ak_y k + \omega t))$ we get the dispersion relation.

2.10: The number of frequencies is N. Thus $\int_0^{\omega_m} g(\omega)d\omega = N$. Use the fact that $g(\omega)d\omega = Cdk$ and hence $g(\omega) = C/|d\omega/dk| = (2C)/(a\sqrt{\omega_m^2 - \omega^2})$ where $\omega_m = 2\sqrt{F/m}$ (the maximum frequency). The normalization determines C and hence we get

$$g(\omega) = \frac{2N}{a}\frac{1}{\sqrt{\omega_m^2 - \omega^2}} \quad \text{(AN.29)}$$

2.11: At 10 K we obtain 0.005 J/Kmol from the electrons and 0.0593 J/Kmol from the phonons.

2.12: The force on a unit cell is gpa^2, which is equal to $4k\Delta r$, where k is the force constant and Δr the change in bond length. Let $\Delta V \sim (a + \Delta r)^3 - a^3$ to get

$$k \sim \frac{3}{4}\frac{ga}{C} \qquad \text{(AN.30)}$$

with $C = 3.110^{-7}$ cm^2/kg, we get $k = 5.6$ amuτ^{-2} where $\tau = 10^{-14}$ sec.

2.13: Use the fact that $k_f \cos\theta_i = K_i + G$ and

$$\omega = \frac{E_i}{\hbar}(k_f^2/k_i^2 - 1) = \frac{E_i}{\hbar}((K_i/k_i + G/k_i)^2/\cos^2\theta_i - 1) \qquad \text{(AN.31)}$$

Finally introduce $G = \Delta K$ and $K_i/k_i = \sin\theta_i$.

2.14: Energy conservation gives

$$\frac{1}{2m}(P_0^g)^2 + \frac{1}{2M_s}(P_0^s)^2 = \frac{1}{2m}(P_1^g)^2 + \frac{1}{2M_s}(P_1^s)^2 \qquad \text{(AN.32)}$$

where P_0^g and P_1^g are the initial and final momenta of the gas atom. Similarly P_0^s and P_1^s are momenta of the surface atom. Momentum conservation gives

$$P_0^g + P_0^s = -P_1^g + P_1^s \qquad \text{(AN.33)}$$

where we have used the fact that the gas momentum changes sign after the collision. Thus we have $P_1^s = P_0^g + P_1^g + P_0^s$ and

$$\Delta E = \frac{1}{2m}((P_0^g)^2 - (P_1^g)^2) = \frac{1}{2M_s}((P_1^s)^2 - (P_0^s)^2) \qquad \text{(AN.34)}$$

Using now the relation $P_1^g \sim P_0^g - (m\Delta E)/(P_0^g)$ we get

$$P_1^s = 2P_0^g + P_0^s - \frac{m\Delta E}{P_0^g} \qquad \text{(AN.35)}$$

which when inserted in the expression for ΔE gives

$$\Delta E = \frac{4}{2M_s}\left((P_0^g)^2 + \frac{m^2}{M_s^2}(P_0^g + P_0^s)^2 - \frac{2mP_0^g}{M_s}(P_0^s + P_0^g) + P_0^s P_0^g - \frac{m}{M_s}P_0^s(P_0^g + P_0^s)\right) \qquad \text{(AN.36)}$$

where we have used the fact that to first order $\Delta E \sim (4P_0^g)/(2M_s)(P_0^g + P_0^s)$. Introducing now a thermal average of the surface momenta such that $\langle P_0^s \rangle = 0$ and $\langle (P_0^s)^2 \rangle = 2M_s k_B T_s$ we get the final result.

We obtain $\alpha = 0.22$ and 0.59 for He and Ne resp. With the Baule formula we have 0.22 and 0.73.

Chapter 3

3.1: Integration over x gives

$$\int_{-\infty}^{\infty} \exp(i\Delta K_x)\exp(-x^2/4Dt)\,dx = \sqrt{4\pi Dt}\exp(-\Delta K_x^2 Dt) \qquad \text{(AN.37)}$$

With a similar result for the y component, we can evalute the integral over t.

3.2: We use the fact that

$$\Delta \mathbf{R} = \int_0^t dt'\mathbf{v}(t') \qquad \text{(AN.38)}$$

and the fact that for the ensemble average we have

$$\left\langle \int_0^t \mathbf{v}(t')\,dt' \int_0^t \mathbf{v}(t'')\,dt'' \right\rangle = \int_0^t dt' \int_0^t dt'' \langle \mathbf{v}(t'-t'')\cdot\mathbf{v}(0)\rangle \qquad \text{(AN.39)}$$

We can now change variable, i.e., introduce $\tau = t' - t''$ and use

$$\begin{aligned} -\int_0^t dt' \int_{t'}^{t'-t} d\tau &= \int_0^t dt' \int_0^{t'} d\tau - \int_0^t dt' \int_0^{t'-t} d\tau \\ &= \int_0^t d\tau \int_\tau^t dt' + \int_0^t d\tau \int_0^{t+\tau} dt' \\ &= \int_0^t d\tau (t-\tau) + \int_0^t (t+\tau)\,d\tau \\ &= 2t \int_0^t d\tau \end{aligned} \qquad \text{(AN.40)}$$

which gives the desired result.

3.3: $D_0 = 2.8\ 10^{-4}$ cm^2/sec and $E_0 = 57$ meV.

3.4: We obtain $E_d = 5.0$ kJ/mol and $D_0 = 8.0\ 10^{-4}$ cm^2/sec from the values $\nu_0 = 1.0, 1.2,$ and $1.10\ 10^{12}$ sec^{-1} at 250, 200, and 140 K, respectively.

Chapter 4

4.1: We need to evaluate the integral

$$\int_{-\infty}^{\infty} dn \int_{-\infty}^{\infty} dm \int_{0}^{\infty} \frac{1}{((\Delta x - an)^2 + (\Delta y - bm)^2 + (z + cl)^2)^3} \tag{AN.41}$$

where $\Delta x = x - x'$, $\Delta y = y - y'$, and (x, y, z) the position of the atom.

Introduce $u = \Delta x - an$, $v = \Delta y - bm$, then straightforward evaluation of the integrals yields

$$\frac{1}{z^3} \frac{B\left(\frac{1}{2}, \frac{5}{2}\right) B\left(\frac{1}{2}, 2\right) B(1, 3)}{abc} \tag{AN.42}$$

which finally gives the desired result.

4.2: The result is obtained using the same procedure as in exercise 4.1. The minimum value for $V(z)$ is $z_0 = (3C_9/C_3)^{1/6}$ and using tha fact that $C_9/C_3 = 2C_{12}/15C_6$ and $r_0 = (2C_{12}/C_6)^{1/6}$ we get $z_0 = r_0(1/5)^{1/6}$. Substituting the value for z_0 in the 9-3 potential we have $V_b = (2/3)\sqrt{1/3}C_3^{3/2}/C_9^{1/2}$. Using the relation between C_9 and C_{12}, C_3, and C_6, that $C_12 = Dr_0^{12}$ and $C_6 = 2Dr_0^6$ we get the relation between V_b and D.

Chapter 5

5.1: Using the commutation relation $[a, a^+] = 1$ we get the following equations

$$\frac{d}{dt}\begin{bmatrix} O_1 \\ O_2 \\ O_3 \end{bmatrix} = \begin{bmatrix} A_1 & 0 & 0 \\ 0 & A_2 & 0 \\ if(t)/\hbar & -if(t)/\hbar & -2\gamma_2 \end{bmatrix} \begin{bmatrix} O_1 \\ O_2 \\ O_3 \end{bmatrix} + \begin{bmatrix} -if(t)/\hbar \\ if(t)/\hbar \\ 0 \end{bmatrix} \tag{AN.43}$$

with the solution for $O_i = a, a^+$ and a^+a $(i = 1, 2, 3)$

$$O_1 = -\frac{i}{\hbar}\exp\left(\int A_1\, dt'\right) \int f(t')\exp\left(-\int A_1\, dt''\right) dt' \tag{AN.44}$$

$$O_2 = \frac{i}{\hbar}\exp\left(\int A_2\, dt'\right) \int f(t')\exp\left(-\int A_2\, dt''\right) dt' \tag{AN.45}$$

$$O_3 = \frac{i}{\hbar}\exp\left(-2\int \gamma_2\, dt'\right) \int f(t')(O_1(t') - O_2(t')) \times \exp\left(2\int \gamma_2\, dt''\right) dt' \tag{AN.46}$$

5.2: For small values of ε compared with $E_{kin} = \hbar^2 k_i^2/2m$ we have

$$k_f \sim k_i + m\varepsilon/(\hbar^2 k_i) \tag{AN.47}$$

With this approximation, we get the maximum of the intensity as

$$\varepsilon_{max} = \left(E_{int} + \frac{m}{\hbar^2}\sin^2\theta_f \left(\frac{\sigma_0^2}{\sigma_x^2}\cos^2\phi_f \left(\frac{\sin\theta_i \cos\phi_i}{\sin\theta_f \cos\phi_f} - 1 \right) \right.\right.$$
$$\left.\left. + \frac{\sigma_0^2}{\sigma_y^2}\sin^2\phi_f \left(\frac{\sin\theta_i \sin\phi_i}{\sin\theta_f \sin\phi_f} - 1 \right) \right) \right)$$
$$\Big/ \left(1 + \frac{m^2 \sin^2\theta_f}{\hbar^4 k_i^2} \left(\frac{\sigma_0^2}{\sigma_x^2}\cos^2\phi_f + \frac{\sigma_0^2}{\sigma_y^2}\sin^2\phi_f \right) \right) \tag{AN.48}$$

Assuming now $\phi_i = \phi_f$, $\sigma_x = \sigma_y$ and using eqs. (5.43) and (5.46) together with $v_R = a\omega$ we get the following simplified expression

$$\varepsilon_{max} = \frac{E_{int} + m v_R^2 \left(\sin^2\theta_f \left(\dfrac{\sin\theta_i}{\sin\theta_f} - 1 \right) \right)}{1 + \dfrac{0.5 m v_R^2}{E_{kin}} \sin^2\theta_f} \tag{AN.49}$$

We see that we get a negative shift if $\theta_i < \theta_f$ and positive for $\theta_i > \theta_f$. With $\omega = 50¡m^{-1}$ we have a shift of +3.5 meV for $\theta_i = 50.9°$, −4.0 meV for 44.9° and −8.1 and −12.5 for the other two angles.

Chapter 6

6.1: $r_- = 0.6281$ Å and $r_+ = 0.8863$ Å.

6.2: We have $(p_r^2)/(2m) = E - (1/2)k(r - r_e)^2 - (\hbar^2 j(j+1))/(2mr_e^2)$ and

$$(v + 1/2) = \frac{2}{h}\int_{r_-}^{r_+} dr p_r \tag{AN.50}$$

where $r_\pm$ are the classical turning points. Expanding $1/r^2 = 1/r_e^2(1 - 2(r - r_e)/r_e + 3(r - r_e)^2/r_e^2)$ we get $a = 2m\left(E - (\hbar^2 j(j+1))/(2mr_e^2)\right)$, $b = 2j(j+1)\,\hbar^2/mr_e^3$, and $c = -m(k + 3j(j+1)\hbar^2/(mr_e^4))$.

Using that arcsin $(\pm 1) = \pm\pi/2$ and $k = m\omega^2$ we get

$$E = \hbar\omega\left(v + \frac{1}{2}\right) + \frac{\hbar^2 j(j+1)}{mr_e^2} - D_e j^2(j+1)^2 - \alpha_e\left(v + \frac{1}{2}\right) j(j+1) \tag{AN.51}$$

where $D_e = \hbar^4/(2m^3\omega^2 r_e^6)$ and $\alpha_e = -3\hbar^3/(2m^2\omega r_e^4)$. For $\beta r_e << 1$ we have $D_e\beta^2 = k/2 = m\omega^2/2$ and hence the expressions for the Morse oscillator become identical to those for the harmonic oscillator.

Chapter 7

7.1: Introduce the relations

$$\dot{\eta} = \mathbf{T}\dot{\mathbf{Q}}(t) \tag{AN.52}$$

$$\dot{\eta}(0)^+ = \dot{\mathbf{Q}}^+(0)\mathbf{T}^+ \tag{AN.54}$$

and the fact that the elements of the matrix $M = \dot{\mathbf{Q}}(t)\dot{\mathbf{Q}}^+(0)$ are given by

$$2\sqrt{E_k E_l}\sin(\omega_k t + \phi_k^0)\sin(\phi_l^0) \tag{AN.53}$$

The ensemble average gives

$$< \sin\phi_k^0 \sin\phi_l^0 > = \frac{1}{2}\delta_{kl} \tag{AN.55}$$

$$< \cos\phi_k^0 \sin\phi_l^0 > = 0 \tag{AN.56}$$

and hence the matrix $< M >$ is diagonal with elements $E_k\cos(\omega_k t)$. Using that

$$\int_0^\infty \cos(\omega_k t)\cos(\omega t) \sim \frac{1}{2}\delta(\omega_k - \omega) \tag{AN.57}$$

we see that the time-integration projects the frequencies one by one and adds unity to $g(\omega)$ (since $\mathbf{TT}^+ = \mathbf{I}$) for each frequency in the spectrum found equal to ω (q.e.d).

7.2: Use that fact that

$$\begin{aligned}\int_0^\infty \int_0^t dt' &= \int_0^\infty dt' \int_{t'}^\infty dt \\ &= \int_0^\infty P(t') \int_0^\infty du L(u)\cos(\omega(u + t')) \\ &= P_c\Lambda_c - P_s\Lambda_s \end{aligned} \tag{AN.58}$$

Thus from the cos-transform one obtains using partial integration

$$P(0) = Ps\left(\frac{B_{11}^*}{\omega} - \omega - \Lambda_s\right) + P_c\Lambda_c \qquad \text{(AN.59)}$$

where

$$P_c = \int_0^\infty P(t)\cos\omega t \qquad \text{(AN.60)}$$

$$P_s = \int_0^\infty P(t)\sin\omega t \qquad \text{(AN.61)}$$

From the sin-transform we get

$$0 = P_s\Lambda_c + P_c\left(\Lambda_s + \omega - \frac{B_{11}^*}{\omega}\right) \qquad \text{(AN.62)}$$

From these two equations, one obtains the final result.

7.3: Use the fact that the Laplace transform of Q_k can by partial integration be written as

$$L(Q_k) = \int_0^\infty dt\exp(-st)Q_k = \frac{1}{s}Q_k^0 + \frac{1}{s^2}\dot{Q}_k^0 + \frac{1}{s^2}L(\ddot{Q}_k) \qquad \text{(AN.63)}$$

and

$$L(\ddot{Q}_k) = -\omega_k^2 L(Q_k) - c_k(L(z) - z_{eq}/s) \qquad \text{(AN.64)}$$

Solving for $L(Q_k)$ we get eq. (7.35). For the inverse transform we use the fact that

$$L^{-1}(s/(\omega_k^2 + s^2)) = \cos(\omega_k t) \qquad \text{(AN.65)}$$

$$L^{-1}(\omega_k/(\omega_k^2 + s^2)) = \sin(\omega_k t) \qquad \text{(AN.66)}$$

$$L^{-1}(A/s) = A \qquad \text{(AN.67)}$$

7.4: Use the fact that from the equipartition theorem we have $Q_k^0 = \sqrt{k_B T_s}/\omega_k\cos\phi_k^0$ where ϕ_k^0 is a random phase angle. Hence we have $< Q_k^0 Q_{k'}^0 >= k_B T_s \delta_{kk'}/\omega_k^2$. Thus we get $< f(0)f(t) >= k_B T_s \sum_k (c_k^2)/(\omega_k^2)\cos\omega_k t$. All other terms vanish. The Laplace transform of $< f(0)f(t) >$ gives

$$k_B T_s \sum_k \frac{c_k^2}{\omega_k^2}\frac{s}{s^2 + \omega_k^2} \qquad \text{(AN.68)}$$

The sum over k was evaluated to ηm and hence the inverse Laplace transform gives the desired result, since $L^{-1}(1) = \delta(t)$.

7.5: Straightforward integration of the Langevin equation gives

$$\mathbf{v}(t) = \exp(-\eta t)\mathbf{v}(0) + \mathbf{g}(t) \tag{AN.69}$$

where $\mathbf{g}(t)$ is a fluctuating term. Thus we get

$$D = \frac{1}{n_d} \int d\tau < (\exp(-\eta t)\mathbf{v}(0) + \mathbf{g}(t)) \cdot \mathbf{v}(0) >= \frac{k_B T}{m\eta} \tag{AN.70}$$

where we have used the fact that the ensemble average over the fluctuationg term is zero and that according to the equipartition theorem $< \mathbf{v}(0) \cdot \mathbf{v}(0) >= n_d k_B T/m$.

Chapter 8

8.1: The trajectory is obtained from the integral

$$\int_{z_0}^{z} \frac{dz}{\sqrt{2(E_0 - V(z))/m}} = t - t_0 = t \tag{AN.71}$$

which gives [285]

$$\beta t\sqrt{a} = \ln\left(\frac{(2a + by + 2\sqrt{aR})y_0}{y(2a + by_0 + 2\sqrt{aR_0})}\right) \tag{AN.72}$$

where $y = \exp(-\beta z)$, $a = A/E_0$, $b = B/E_0$, $R = a + by + cy^2$, $y_0 = \exp(-\beta z_0)$, and $R_0 = a + by_0 + cy_0^2$. Using now that $R_0 = 0$ we get

$$y/y_0 = 2/((2 + by_0)\cosh(at) - by_0) \tag{AN.73}$$

where $a = \beta v_0 \cos\theta$.

The forcing integral can be evaluated using the fact that [285]

$$\int_0^\infty \frac{\cos(\omega t)}{(\cosh(at) - \Delta)^2} = \frac{\pi}{a} \frac{\Delta \sinh(\omega\tau/a) + \frac{\omega}{a} \sin\tau \cosh(\omega\tau/a)}{\sin^3 \tau \sinh(\omega\pi/a)} \tag{AN.74}$$

$$\int_0^\infty \frac{\cos(\omega t)}{\cosh(at) - \Delta} = \frac{\pi}{a} \frac{\sinh(\omega\tau/a)}{\sin\tau \sinh(\omega\tau/a)} \tag{AN.75}$$

where $\cos\tau = -\Delta$.

8.2: Using the stochastic probabilities we obtain the following expression for ΔE_k^+

$$\Delta E_k^+ = \hbar\omega_k(1 - z_k)$$
$$\sum_{m=0}^{\infty}\sum_{n_k^0=0}^{\infty} z_k^{n_k^0}\frac{m}{\rho_k}\exp(-(2n_k^0 + 1 + m)/\rho_k)$$
$$I_0\left(2\sqrt{(n_k^0 + 1/2)(n_k^+ + m + 1/2)}/\rho_k\right) \qquad \text{(AN.76)}$$

Introducing $x = n_k^0 + (1/2)$ we obtain for the integral of x

$$z_k^{-1/2}\frac{1}{\sqrt{a_k}}\exp\left(\frac{m}{2}(\beta\hbar\omega_k - \sqrt{a_k})\right) \qquad \text{(AN.77)}$$

Finally the integration over m gives the desired result.

8.3: The displacement of atom α in direction γ is given as

$$r_{\alpha\gamma} - r_{\alpha\gamma}^0 = \frac{1}{\sqrt{m_\alpha}}\sum_{k=1}^{M} T_{\alpha\gamma;k}Q_k \qquad \text{(AN.78)}$$

where Q_k is the normal mode coordinate. Thus using eq. (6.16) we have

$$< \cos(\omega_k t + \phi_k^0)\cos(\omega_{k'} t + \phi_{k'}^0) >\sim \frac{1}{2}\delta_{kk'} \qquad \text{(AN.79)}$$

where an average over one vibrational period has been assumed. This gives

$$(\mathbf{u}\cdot\Delta\mathbf{k})^2 \sim \frac{1}{m_\alpha}\sum_{k\gamma\gamma'}\Delta k_\gamma \Delta k_{\gamma'}\frac{E_k}{\omega_k^2}T_{\alpha\gamma;k}T_{\alpha\gamma';k} \qquad \text{(AN.80)}$$

Introducing the average energy for the phonons

$$\langle E_k\rangle = \frac{\hbar\omega_k}{1 - z_k} \qquad \text{(AN.81)}$$

and using that $\sum_k T_{\alpha\gamma;k}T_{\alpha\gamma';k} = \delta_{\gamma\gamma'}$, also introducing a Debye distribution

$$g(\omega) = \frac{3\omega^2}{\omega_D^3} \qquad \text{(AN.82)}$$

we can replace the summation over k with an integral $\int g(\omega)d\omega$. With these approximations we get

$$\frac{1}{m_\alpha}\sum_\gamma \Delta k_\gamma^2 \frac{3}{\omega_D^3}\int_0^{\omega_D} d\omega \frac{\hbar\omega}{1-\exp(-\hbar\omega/k_BT)} \tag{AN.83}$$

which if the z exchange is the dominating term in the sum and $\hbar\omega/k_BT << 1$ gives

$$\frac{3\hbar^2\Delta k_z^2 T}{m_\alpha k_B \Theta_D^2} \tag{AN.84}$$

where we have used the fact that $\Theta_D = \hbar\omega_D/k_B$.

8.4: The D - W factor is given by

$$2W = \frac{8\pi^2 m^2}{\hbar\alpha^2 M}\int_0^{\omega_D} d\omega \frac{3\omega^3}{\omega_D^3}\exp(-2\pi\omega/\alpha v_0)\coth(\hbar\omega/2kT) \tag{AN.85}$$

which can be reformulated to

$$2W = \frac{36\alpha^2 E_0^2}{\pi^2\hbar M\omega_D^3}F(z) \tag{AN.86}$$

where $z = 2\pi kT/(\alpha v_0\hbar)$ and

$$F(z) = \frac{(2z)^4}{\Gamma(4)}\int_0^{x_D} dx x^3 \exp(-2zx)\coth(x) \tag{AN.87}$$

with $x_D = \hbar\omega_D/2kT$. For $z = 0$ we have $F(0) = 0$, i.e., $W = 0$ and for large values of x_D we can evaluate the integral

$$F(z) \sim 2z^4\zeta(4, z) - 1 \tag{AN.88}$$

where $\zeta(4, z)$ is Riemann's Zeta function [284], i.e.,

$$\zeta(q, z) = \sum_{n=0}^{\infty}\frac{1}{(n+z)^q} \tag{AN.89}$$

Using this expression we get $W(200\ K) = 2.2$ and $W(300K) = 3.2$, i.e., roughly proportional to T_s as also indicated by eq. (8.127).

With the expression (8.128) we obtain 2W = 3.66 at 200 K and 5.49 at 300 K. In order to obtain the same value we must assume $\alpha \sim 4.5\mathring{A}^{-1}$.

8.5: $\langle n \rangle$ is given by

$$\langle n \rangle = \sum_{n=0}^{\infty} n P_{0 \to n} = \sum_{n=0} n \frac{\rho^n}{n!} \exp(-\rho) = \rho \tag{AN.90}$$

Chapter 9

9.1: Using eq. (9.35) and that the density N/V is 0.0849 electrons / Å^3 we get $k_F = 1.36$ Å^{-1} and $\varepsilon_F = 680$ kJ/mol or a Fermi-temperature of 82000 K.

9.2: The potential created by the charge distribution $f(\mathbf{r}) = \exp(i(\mathbf{k} - \mathbf{k}') \cdot \mathbf{r})$ is

$$\phi(\mathbf{r}') = \int \frac{f(\mathbf{r})}{|\mathbf{r}|' - \mathbf{r}} d\mathbf{r} \tag{AN.91}$$

and hence using the Poisson equation, we get

$$\phi(\mathbf{r}) = \frac{4\pi \exp(i(\mathbf{k} - \mathbf{k}') \cdot \mathbf{r})}{|\mathbf{k} - \mathbf{k}'|^2} \tag{AN.92}$$

We can rewrite the integral

$$\int \frac{d\mathbf{k}'}{|\mathbf{k} - \mathbf{k}'|^2} \tag{AN.93}$$

as

$$\int_0^{k_0} dk' \int_0^{\pi} d\theta \int_0^{2\pi} d\phi \frac{k'^2 \sin\theta d\phi}{k^2 + k'^2 - 2kk' \sin\theta} \tag{AN.94}$$

giving

$$\frac{\pi k_0}{2}[2 + \frac{k_0^2 - k^2}{k k_0} \ln|\frac{k_0 + k}{k_0 - k}|] \tag{AN.95}$$

9.3: The energy levels are given by

$$\varepsilon_N = \frac{\pi^2 \hbar^2 (n_x^2 + n_y^2 + n_z^2)}{2L^2 m} \tag{AN.96}$$

where $N^2 = n_x^2 + n_y^2 + n_z^2$. The number of states in 1/8 (since $n_x > 0$, etc.) of a sphere with radius n_{max} is proportional to the volume, i.e.,

$$N = \frac{1}{8}(4\pi n_{max}^3)/3 \tag{AN.97}$$

The density of states is $n(\mathbf{r}) = 2N/V$, where we have operated with doubly occupied states. Or since $n_{max} = (6N/\pi)^{1/3}$, we have

$$n(\mathbf{r}) = \frac{\pi n_{max}^3}{3V} \tag{AN.98}$$

The energy can be obtained as

$$E_k = \sum_{N<n_{max}} \varepsilon_N = 2\int_0^{n_{max}} dN \frac{1}{8} 4\pi N^2 \frac{\pi^2 N^2 \hbar^2}{2L^2 m} \tag{AN.99}$$

or

$$E_k = \frac{3}{10} V \frac{\hbar^2 (3\pi^2)^{2/3}}{m} n(\mathbf{r})^{5/3} \tag{AN.100}$$

where we have used $L^2 = V^{2/3}$ and (AN.98). Since the density is uniform this may be written as

$$E_k = \frac{3\hbar^2}{10m}(3\pi^2)^{2/3} \int n(\mathbf{r})^{5/3} d\mathbf{r} \tag{AN.101}$$

q.e.d.

9.4: Using $n = n_0$ and $v_0 = \hbar k_F/m$ we get the screening length $\sqrt{0.75}\sqrt{\pi a_0/k_F}$ in the plasma model compared with $0.5\sqrt{\pi a_0/k_F}$ in the Thomas-Fermi model, i.e., a difference of a factor 1.73.

Chapter 10

10.1: For small values of t we have $H \rightarrow (e^2/a_0)\beta\phi^3 t^2$ and the opposite limit gives $H \rightarrow -\varepsilon_c^{unif}$.

Chapter 11

11.1: Consider the equation for the adiabatic states

$$H_{el}\phi_m(\mathbf{r}, \mathbf{R}) = E_m(\mathbf{R})\phi_m(\mathbf{r}, \mathbf{R}) \tag{AN.102}$$

If we operate from the left with $\nabla_{\mathbf{R}}$ and then ϕ_n^* we get

$$\langle\phi_n|\nabla H_{el}|\phi_m\rangle + \langle\phi_n|H_{el}|\nabla\phi_m\rangle = \nabla E_m\delta_{nm} + E_m\langle\phi n|\nabla\phi_m\rangle \quad \text{(AN.103)}$$

and using that fact the second term is $E_n\langle\phi_n|\nabla\phi_m\rangle$ we get the result.

11.2: We find $N(E)$ as

$$N(E) = 4\frac{2}{AL_z}\sum_{n_x}\sum_{n_y}\sum_{n_z}\sin^2(k_z z) \quad \text{(AN.104)}$$

Converting to an integral we have

$$N(E) = 4\frac{2}{AL_z}\int dn_x\int dn_y\int_0 dn_z\sin^2(k_z z) \quad \text{(AN.105)}$$

Introducing now

$$E = \frac{h^2}{2m}\left(\frac{n_x^2}{L_x^2} + \frac{n_y^2}{L_y^2} + \frac{n_z^2}{L_z^2}\right) \quad \text{(AN.106)}$$

$A = 4L_x^2 = 4L_y^2$, $n_x = n\cos\phi$, and $n_y = n\sin\phi$, we can rewrite the integral as

$$N(E) = 4\frac{4\pi}{AL_z}\int_0^{n_{max}} dnn\left(x/2 - \frac{1}{4a}\sin(2ax)\right) \quad \text{(AN.107)}$$

where $a = 2\pi z/L_z$, $x = \sqrt{8mE/h^2}\sqrt{1-y}$, and $y = 2h^2n^2/mEA$. The integration is changed to one over y instead of n with the result that

$$N(E) = \frac{Em}{2h^2 z}\left(\frac{2}{3}B - \frac{2}{B^2}\sin B + \frac{2}{B}\cos B\right) \quad \text{(AN.108)}$$

where $B = 2z\sqrt{2mE}/\hbar$. From $dN(E)/dE$ we finally obtain $n(E, z)$.

11.3: Interchange of say particle i and k can be performed with, for instance, the operator $c_k^+c_jc_i^+c_kc_j^+c_i$, where $n_j = 0$ and $n_i = n_k = 1$. Assume that $i < k < j$. The total phasefactor is $\Sigma_{p<i}n_p + \Sigma_{p<j}n_p + \Sigma_{p<k}n_p + \Sigma_{p<i}n_p + \Sigma_{p<j}n_p + \Sigma_{p<k}n_p$. The two phasefactors $\Sigma_{p<i}$ cancel since states below i are not affected by the operation. Likewise for $\Sigma_{p<j}n_p$ since the total number of quanta below j is one lower than originally, both times the c_j or c_j^+ operators are used. However, when the c_k operator is used, the number

of quanta less than k is one smaller than when c_k^+ is used. Hence a factor of (-1) appears.

11.4: In both cases we get $O_{il}\delta_{jk} - O_{kj}\delta_{il}$.

11.5: Using the above result we get

$$O_{ik}O_{kk}^p = (O_{kk} + 1)^p O_{ik} \tag{AN.109}$$

$$O_{kj}O_{kk}^p = (O_{kk} - 1)^p O_{kj} \tag{AN.110}$$

and $O_{ij}O_{kk} = O_{kk}O_{ij}$ if $k \neq j$ or i. Using these results we get that

$$U_0^+ O_{ij} U_0 = O_{ij} \exp\left(\frac{i}{\hbar}(\varepsilon_i - \varepsilon_j)t\right) \tag{AN.111}$$

where we have used the fact that $U_0 = \exp\left((1)/(i\hbar)\sum_k \varepsilon_k t\right)$.

INDEX

DATE DUE
